D0984107

CELL–CELL SIGNALING IN VERTEBRATE DEVELOPMENT

CELL–CELL SIGNALING IN VERTEBRATE DEVELOPMENT

Edited by

ELIZABETH J. ROBERTSON
FREDERICK R. MAXFIELD
HENRY J. VOGEL

College of Physicians and Surgeons
Columbia University
New York, New York

ACADEMIC PRESS, INC.
A Division of Harcourt Brace & Company

San Diego   New York   Boston
London   Sydney   Tokyo   Toronto

This book is printed on acid-free paper. ∞

Academic Press, Inc.
1250 Sixth Avenue, San Diego, California 92101-4311

United Kingdom Edition published by
Academic Press Limited
24–28 Oval Road, London NW1 7DX

Library of Congress Cataloging-in-Publication Data

Cell-cell signaling in vertebrate development / edited by Elizabeth J.
   Robertson, Frederick R. Maxfield, Henry J. Vogel.
        p.   cm.
     Includes index.
     ISBN 0-12-590370-7
     1. Vertebrates--Development.   2. Cell interaction.
   3. Vertebrates--Cytology.   4. Developmental cytology.
   I. Robertson, E. J. (Elizabeth J.)   II. Maxfield, Frederick R.
   III. Vogel, Henry J. (Henry James), Date
   QL959.C35   1993
   596'.03'33--dc20   93-7463
                                                          CIP

PRINTED IN THE UNITED STATES OF AMERICA
93   94   95   96   97   98   BC   9   8   7   6   5   4   3   2   1

Contents

3   Mesoderm Induction and Pattern Formation in the Amphibian Embryo

IGOR B. DAWID, MICHAEL R. REBAGLIATI, AND MASANORI TAIRA

4   Relationships between Mesoderm Induction and Formation of the Embryonic Axis in the Chick Embryo

CLAUDIO D. STERN

PART III   CELL MIGRATION AND DIFFERENTIATION

5   Studies of Neural Cell Lineages Using Injectable Fluorescent Tracers

RICHARD WETTS AND SCOTT E. FRASER

6   Axon Guidance in the Mammalian Spinal Cord

JANE DODD AND THOMAS M. JESSELL

7   Axon Patterning in the Visual System: Divergence of Retinal Axons to Each Side of the Brain at the Midline of the Optic Chiasm

CAROL A. MASON AND PIERRE GODEMENT

PART V   PATTERN FORMATION

12   The Relationship between *Krox-20* Gene Expression and the Segmentation of the Vertebrate Hindbrain

M. ANGELA NIETO, LEILA C. BRADLEY, AND
DAVID G. WILKINSON

13   The Effect of Retinoids on Amphibian Limb Regeneration

JEREMY P. BROCKES

PART VI   TRANSCRIPTION FACTORS

14   Role of Transcription Factor GATA-1 in the Differentiation of Hemopoietic Cells

LARYSA PEVNY, M. CELESTE SIMON, VIVETTE D'AGATI,
STUART H. ORKIN, AND FRANK COSTANTINI

Preface

How cells become specified and differentiate in the course of development constitutes a central theme of developmental biology. A key process leading to cell-type differences and to cell organization in vertebrates is embryonic induction in which the developmental fate of a cell is influenced by its proximity to a cell or cells of another type. The elucidation of this process, in molecular terms, represents a long-standing problem to which molecular biological approaches have recently been applied. Involvement of differential gene expression, controlled by transcription factors specifically expressed during development, is being recognized as part of a widespread basic mechanism. There is a growing realization that genes concerned with cell or tissue interactions can be members of multigene families having common functional domains or overlapping functions. Likewise, families of transcription factors, as well as of growth factors and their receptors, have been implicated in inductive processes and patterning.

This volume focuses on developmentally significant cell–cell interactions and on the molecular signals that mediate them. It is arranged according to major developmental phenomena demonstrated in illustrative systems derived from amphibian, avian, mammalian, and piscine sources. Part I introduces the topic of gene activation in the context of early vertebrate development. Part II is concerned with cellular contacts and the induction process. Types of cell–cell interactions are illustrated in analyses of neurogenesis in the mouse; embryonic induction, particularly that of mesoderm, is considered in the frog and in the chick. Part III, dealing with cell migration, begins with a discussion of cell lineages in the frog eyebud. Then, migration phenomena are presented in connection with axon guidance in the embryonic rat spinal cord and mouse visual system. Last, there is a chapter on pathfinding by primary motoneurons, followed by the formation of terminal arbors in zebrafish embryos. Part IV pertains to developmental processes thought to depend on diffusible signals and signal gradients. This section includes a discussion of a mouse gene family that encodes (highly conserved) proteins mediating tissue interactions in embryonic development; inductive effects of signal gradients are considered for the chick embryo with respect to neural-cell identity and limb development. Part V illustrates pattern formation as exemplified in the developing chick hindbrain and in urodele limb regeneration. Part VI highlights gene expression and its regulation by transcription factors or growth factors in rodent development. Two gene families are discussed: homeobox-containing genes, in relation to regional diversity in the head, and paired box-containing genes, in relation to the central nervous system. The role of a member of the GATA family of transcription factors is described, as is the inducing role of nerve growth factor. These two factors participate in erythroid and neuronal differentiation, respectively.

It is hoped that this volume reflects the excitement being generated in this highly active field. Accelerating advances are clearly foreshadowed—advances in the direction of understanding the precise coordination of the growth and differentiation of developing tissues.

It is a pleasure to acknowledge the advice and help of Dr. F. D. Costantini, Dr. A. Efstratiadis, Dr. P. Gruss, Dr. J. B. Gurdon, Dr. M. E. Hatten, Dr. T. M. Jessell, Dr. C. Mason, Dr. V. Papaioannou, and Dr. E. B. Ziff.

We are grateful for the continued interest of Dr. Donald F. Tapley and for the fine support of the College of Physicians and Surgeons (P&S) of Columbia University without which this volume would not have reached fruition. This volume was developed from a P&S Biomedical Sciences Symposium held at Arden House on the Harriman Campus of Columbia University.

Elizabeth J. Robertson*
Frederick R. Maxfield
Henry J. Vogel

---

*Present address: Departments of Cellular and Developmental Biology, and Biochemistry and Molecular Biology, Harvard University, Cambridge, Massachusetts 02138.

PART I

INTRODUCTION

1

Mechanisms of Gene Activation in Early Vertebrate Development

J.B. GURDON,[1] N.D. HOPWOOD, AND M.V. TAYLOR
Wellcome CRC Institute, University Cambridge, Tennis Court Road,
Cambridge CB2 1QR, United Kingdom

The single most important mechanism which leads to cell-type differences and cell organization in the vertebrate animals is embryonic induction. This term refers to any situation in which the developmental fate of a cell is redirected by its proximity to a cell or cells of another type. The phenomenon was first discovered in 1901 and rigorously documented in 1924. It is now understood that nearly all the first cell-type differences in amphibian development arise by a sequential series of embryonic inductions. The first of these, and the one most fully analyzed so far, is that in which cells of the vegetal hemisphere of the blastula redirect cells from the animal hemisphere from an epidermal fate into a mesodermal, and in particular muscle, fate.

Recently there has been much interest in the finding that transforming growth factor (TGF)-β-like substances act as effective inducers. One of these substances, namely activin, is able to generate many different kinds of cells when applied to amphibian blastula cells of the epidermal lineage. Activin is a good candidate for a natural inducer that may be involved at some stage of early amphibian development.

Our approach to understanding the formation of amphibian mesoderm by embryonic induction has been to identify an early specific

---

[1] Based on the opening address given at the P & S Biomedical Sciences Symposium, "Cell–Cell Signaling in Vertebrate Development."

gene response to induction and to try to work backward from there to the first inductive events.

Of the muscle-specific genes whose products accumulate progressively during myogenesis (terminal muscle genes), the muscle-specific actin genes are expressed particularly early. The actin genes which are ultimately expressed only in adult heart and adult skeletal muscle (cardiac and skeletal actin genes) are first transcribed at the mid-gastrula stage of embryos (12 hr from fertilization), even though contracting muscle is not functional until the tail bud stage (about 30 hr from fertilization). The adult cardiac actin gene is expressed in the embryonic axial muscle, before any heart rudiment has been formed. We have chosen to make a detailed analysis of the cardiac actin muscle gene expressed in the axial muscle of embryos. As an experimental induction assay, we combine pieces of ectoderm and endoderm from blastula embryos to observe the transcriptional activation of the cardiac actin gene 9 hr after the inducing endoderm and responding ectoderm tissues have been placed in contact.

Analyzing cardiac actin gene activation at the cellular level, we have found, in addition to the initial receipt of inducer substances by responding cells, a calcium-dependent process, which appears to be an interaction among responding cells with themselves, this process depending on calcium and therefore presumably cell contacts. We have termed this interaction at this stage as "a community effect." We find that a cell can proceed to full muscle differentiation only if it is in close contact with other cells responding to the induction in the same way. Single cells will not normally enter the pathway of muscle differentiation, but many such cells placed together can generate a muscle differentiation response. We think this process may have to do with the release of muscle differentiation factors by cells which have undergone a preliminary response to induction. We suppose that a high concentration of inducer substances is required for complete myogenesis. We have also analyzed the mechanisms by which the spread of the inductive influence is limited to a certain proportion of the responding cells, and find that this depends on several factors of which the slow spread of inducer and the loss of competence are important.

At the molecular level, the requirement for protein synthesis shows that new gene transcription or at least message translation must take place between the receipt of inducer substances and the first transcription of the cardiac actin gene. Using low stringency screening of cDNA libraries, we have found two genes that are involved in the formation of, and patterning within, the mesoderm. One of these is

XMyoD which appears to be closely related to the mammalian *MyoD*, the myogenic gene first described. The *Xenopus MyoD* gene is first transcribed 2 hr before the cardiac actin gene and is already restricted in its expression to that part of the mesoderm which will form muscle. The other *Xenopus* gene relevant to mesoderm patterning is *Xtwi*, found by screening a library with the *Drosophila* mesoderm expressing gene *twist*. *Xtwi* is first expressed at about the same time as cardiac actin but in only those parts of embryonic mesoderm which will not form muscle, i.e., in the notochord and lateral plate mesoderm.

The overexpression of *MyoD* is able to cause blastula cells to be redirected toward muscle differentiation rather than to follow their normal fate and become epidermis. We therefore believe that *MyoD* expression is a normal stage in the process by which cells respond to embryonic induction.

Related to *MyoD* is another myogenic gene termed X-*myf5*. We believe this is the only other myogenic determinant gene which is expressed very early in amphibian development and which may be involved in the initial commitment of cells to a muscle differentiation pathway.

We hope eventually to extend our understanding of the mechanism of activation of *MyoD* and *myf5* and of their action on muscle-specific actin. In this way it should eventually be possible to complete the understanding of how an inducer, like activin, is able to selectively initiate the transcription of genes like *MyoD* and *myf5* in the future muscle cell lineage.

REFERENCES

Gurdon, J. B. (1987). Embryonic induction—molecular prospects. *Development* **99**, 285–306.

Gurdon, J. B. (1988). A community effect in animal development. *Nature* **336**, 772–774.

Gurdon, J. B. (1989). The localization of an inductive response. *Development* **105**, 27–33.

Gurdon, J. B., Mohun, T. J., Taylor, M. V., and Sharpe, C. R. (1989). Embryonic induction and muscle gene activation. *Trends Genet.* **5**(2), 51–56.

Gurdon, J. B. (1989). From egg to embryo: The initiation of cell differentiation in amphibia. *Proc. R. Soc. London B.* **237**, 11–25.

Gurdon, J. B., Tiller, E., Roberts, J., and Kato, K. (1993). A community effect in muscle development. *Curr. Biol.* **3**, 1–11.

Hopwood, N. D., Pluck, A., and Gurdon, J. B. (1989). MyoD expression in the forming somites is an early response to mesoderm induction in *Xenopus* embryos. *EMBO J.* **8**(11), 3409–3417.

Hopwood, N. D., Pluck, A., and Gurdon, J. B. (1989). A *Xenopus* mRNA related to *Drosophila twist* is expressed in response to induction in the mesoderm and the neural crest. *Cell* **59**, 893–903.

Hopwood, N. D., and Gurdon, J. B. (1990). Activation of muscle genes without myogenesis by ectopic expression of MyoD in frog embryo cells. *Nature* **347**, 197–200.

Hopwood, N. D., Pluck, A., and Gurdon, J. B. (1991). *Xenopus* Myf-5 marks early muscle cells and can activate muscle genes ectopically in early embryos. *Development* **111**, 551–560

Mohun, T. J., Taylor, M. V., Garrett, N., and Gurdon, J. B. (1989). The CArG promoter sequence is necessary for muscle-specific transcription of the cardiac actin gene in *Xenopus* embryos. *EMBO J.* **8**(4), 1153–1161.

Taylor, M. V., Gurdon, J. B., Hopwood, N. D., Towers, N., and Mohun, T. J. (1991). *Xenopus* embryos contain a somite specific MyoD-like protein which binds to a promoter site required for muscle actin expression. *Genes Dev.* **5**, 1149–1160.

PART II

CELL–CELL CONTACTS AND INDUCTION

2

Control of Granule Cell Neurogenesis and Migration in Developing Cerebellar Cortex

W.-Q. GAO,[1] G. FISHELL,[1] S. KUHAR,* L. FENG,* N. HEINTZ,* AND M. E. HATTEN[1]

Department of Pathology
Center for Neurobiology and Behavior
College of Physicians and Surgeons
Columbia University, New York, New York 10032
*and *Howard Hughes Medical Institute*
The Rockefeller University, New York, New York 10021

INTRODUCTION

In the vertebrate brain, neurogenesis is restricted to discrete zones, the ventricular zones (VZ) along the inner surface of the neural tube (His, 1889; Ramon y Cajal, 1911; Boulder Committee, 1970), and the external germinal layer (EGL) of the cerebellar cortex (Ramon y Cajal, 1889; Miale and Sidman, 1961; Angevine and Sidman, 1961). In these zones, precursor cells undergo rapid cell division forming a multilayered pseudoepithelium from which postmitotic progeny migrate to establish the neuronal layers of cortex (Boulder Committee, 1970; Sidman and Rakic, 1973). The spatially restricted pattern of neurogenesis and migration seen in brain contrasts with that of the peripheral nervous system (PNS), where multipotential precursor cells undergo proliferation and neuronal differentiation as they migrate through the tissue (Patterson, 1978; LeDouarin, 1982; Anderson, 1989).

[1] *Present address:* The Rockefeller University, New York, New York 10021.

9

Cell–Cell Signaling in
Vertebrate Development

The cerebellum has provided an important model for studies on mammalian central nervous system (CNS) neurogenesis, because of its relatively simple laminar structure, small number of types of neurons, and well-established pattern of histogenesis. As described by Ramon y Cajal (1889, 1911), the cerebellum has a unique, displaced proliferative zone, the EGL, which generates the most numerous principal neuron of the cerebellar cortex, the granule cell. Histochemical analyses, electron microscopic studies, and [^{3}H]thymidine incorporation studies (Miale and Sidman, 1961; Fujita *et al.*, 1966; Fujita, 1967; Altman, 1972) indicate that cells within the EGL are organized into two discrete zones, a superficial zone, one to two cells thick containing mitotic figures, and an underlying zone of postmitotic cells. It is in this deeper zone where the first evidence of neuronal differentiation can be seen (Ramon y Cajal, 1889, 1911), the extension of granule cell axons, the parallel fibers, and the inward migration of the cell soma of immature granule cells along the radially aligned processes of the Bergmann glial cells (Rakic, 1971; Edmondson and Hatten, 1987; Gregory *et al.*, 1988; Hatten, 1990).

The spatial distribution of dividing and differentiating cells in the EGL suggests that extrinsic signals regulate the specification of granule cell identity. One approach to identifying the signals that regulate granule cell differentiation is to obtain markers that are specifically expressed in discrete zones of the EGL. Molecular biological studies by Kuhar *et al.* (1993) have identified genes with a pattern of expression that is restricted to the outer layer, the proliferative zone of the EGL. Other analyses indicate that expression of late neuronal markers commences in the deeper layers of the EGL, where cells are exiting the cell cycle and initiating neurite extension. These include the axonal glycoprotein, TAG-1 (Furley *et al.*, 1990), the neuron–glia adhesion system, astrotactin (Edmondson *et al.*,1988; Stitt and Hatten, 1990), and the neuronal intermediate filament protein, α-internexin (Kaplan *et al.*, 1990).

To provide a model system for EGL neurogenesis and neuronal differentiation, we have developed methods to purify EGL precursor cells and examine the contribution of interactions among neuronal precursor cells to the specification of granule cell identity. Three general models could be tested with the *in vitro* system. In one model, neurogenesis would occur by an intrinsic program of gene expression. In this model, neuronal differentiation would proceed "on schedule" when the cells were dispersed in cell culture. In two alternative models, local signals among cells in the EGL would control neurogenesis, either by diffusible signals that might be concentrated in this zone or by membrane-bound signals. To test these possibilities, we cultured purified EGL cells as a dispersed cell monolayer in a three-dimensional collagen gel where the cells would be present at high cell density, yet prevented from forming contacts, or in cellular reaggregates, where the cells would form cell–cell contacts.

GRANULE CELL DNA SYNTHESIS STIMULATED BY CONTACTS AMONG NEURONAL PRECURSOR CELLS

When immature granule cells purified from P5 mouse cerebellum by Percoll gradient separation and subsequent panning steps (Hatten, 1985, 1987) were plated as a dispersed monolayer, low levels of DNA synthesis were detected by [^{3}H]thymidine uptake assays (Fig. 1a). Similar results were obtained when the cells were plated on a variety of adhesive substrata, including polylysine, laminin, or Matrigel (Malvern, PA). To examine whether soluble factors regulated

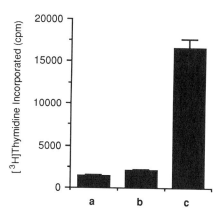

Fig. 1. DNA synthesis of granule cell neuroblasts *in vitro*. [^{3}H]Thymidine incorporation by P5 granule cells maintained for 24 hr *in vitro* in (a) a monolayer (5×10^5 cells/ml), (b) a collagen matrix (1×10^7 cells/ml), and (c) cellular reaggregates (3×10^6 cells/ml). In each case, a total of 5×10^5 cells were plated in serum-free medium. [^{3}H]Thymidine was added 24 hr after plating and incorporation was measured 21 hr later. Data from triplicate cultures are expressed as means ± SEM. Cells in reaggregate cultures show 10-fold higher [^{3}H]thymidine incorporation than cells dispersed in a monolayer or embedded in a collagen matrix. Reprinted with permission from Gao *et al.* (1991).

granule cell proliferation, we plated freshly purified granule cells at high cell density (1×10^7 cells/ml) in a collagen matrix. Under these culture conditions, DNA synthesis levels remained low at levels comparable to those observed for cells dispersed into a monolayer (Fig. 1b). In contrast, cells cultured in large cellular reaggregates (500–2000 cells) showed high levels of DNA synthesis (Fig. 1c). A striking aspect of DNA synthesis in the reaggregates was that it occurred even in the absence of growth factors, suggesting that cell–cell interactions among granule cell progenitor cells stimulated DNA synthesis.

To determine whether the increase in [^{3}H]thymidine incorporation represented dividing cells, we carried out immunocytochemical labeling experiments. By bromodeoxyuridine (BrdU) labeling (Gratzner, 1982), an intensely stained subpopulation of mitotic cells was evident in the cellular reaggregates (Fig. 2). Quantitation of BrdU-labeled cells after 1–4 days *in vitro* by counting random sets of cells in defined optical planes of the reaggregates indicated a mitotic index of 30–40%.

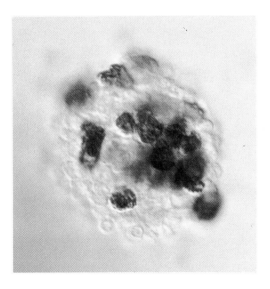

Fig. 2. BrdU labeling reveals mitotic cells in reaggregates. Immunocytochemical localization of BrdU in granule cell neuroblasts in reaggregate cultures. Labeled mitotic cells are evident in reaggregates cultured on untreated culture dishes. After 24 hr *in vitro* culture cells were labeled with BrdU for 21 hr and staining was visualized with Nomarski microscopy. Reprinted with permission from Gao *et al.* (1991).

INFLUENCE OF GROWTH FACTORS ON GRANULE NEUROBLAST DNA SYNTHESIS

To evaluate the contribution of growth factors to granule cell DNA synthesis in reaggregate cultures, we added serum or growth factors to the culture medium. In the presence of serum, granule cell neuroblast DNA synthesis increased 15–20% above the levels seen in serum-free medium (Fig. 3). A larger stimulation, up to twofold, was seen in the presence of insulin-like growth factor 1 (IGF-1), epidermal growth factor (EGF), and basic fibroblast growth factor (bFGF), all of which promoted granule cell DNA synthesis in a dose-dependent manner between 5 and 20 ng/ml. Among the growth factors we tested, IGF-1 was the most potent mitogen. DNA synthesis was not stimulated by nerve growth factor (NGF), platelet-derived growth factor (PDGF), or insulin (5–20 ng/ml).

Growth factors that were mitogenic for immature granule cells (EGF, bFGF, and IGF-1) also stimulated granule cell DNA synthesis

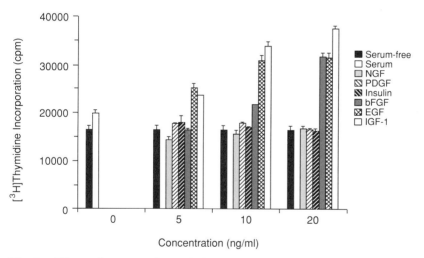

Fig. 3. Effects of serum and growth factors on DNA synthesis. The three sets of histograms (right-hand side) show that the addition of bFGF, EGF, and IGF-1 induces dose-dependent increases in [³H]thymidine incorporation by P5 granule cell neuroblasts cultured in reaggregates over control values in serum-free medium not containing growth factors. In contrast, NGF, PDGF, and insulin did not stimulate granule cell neuroblast DNA synthesis at doses between 5 and 20 ng/ml. The set of histograms (left-hand side) shows that granule cell neuroblast DNA synthesis increased about 20% when cultured in serum-supplemented serum. Data from triplicate cultures are expressed as means ± SEM. Reprinted with permission from Gao *et al.* (1991).

by cells cultured in a monolayer or in a three-dimensional collagen matrix. However, the maximal level of [^{3}H]thymidine incorporation by cells cultured at low cell density (5×10^5 cells/ml) in monolayers in the presence of EGF, bFGF, or IGF-1 (1–20 ng/ml) was 5–10% of the level observed in reaggregates. Higher levels of [^{3}H]thymidine incorporation, up to 25% of the amount seen in cellular reaggregates, were observed when the cells were grown in high density (1×10^7 cells/ml) in a three-dimensional collagen matrix in the presence of IGF-1 (20 ng/ml) (Gao *et al.*, 1991).

DURATION OF PROLIFERATION BY IMMATURE GRANULE CELLS *IN VITRO*

To examine whether the number of cell divisions seen *in vitro* would follow the schedule observed *in vivo* (Angevine and Sidman, 1961; Fujita, 1967; Fujita *et al.*, 1966; Rakic, 1971), we purified immature granule cells at P5, the time when the rate of proliferation of cells in the EGL begins to accelerate, at P8, the peak of cell proliferation, and at P10, a time when the EGL begins to thin and collapse owing to emigration of postmitotic cells, and measured the mitotic index of cells cultured in cellular reaggregates for 1 day to 3 weeks. By BrdU labeling, in all four cases, a similar pattern of labeling was observed. Over the first 4 days *in vitro*, about 30% of the cells were mitotic, and after 10–14 days, approximately 10% of the cells continued to divide in the reaggregate. As cells harvested on P10 would have concluded cell division in 4–5 days *in vivo*, this suggested that signals within the reaggregates, rather than predetermined schedules of proliferation, regulated the proliferative capacity of the cells (Gao *et al.*, 1991).

CHARACTERIZATION OF THE PROGENY OF GRANULE CELL NEUROBLASTS *IN VITRO*

To characterize the progeny of proliferating granule neuroblasts *in vitro*, we carried out light microscopy and immunocytochemical localization of cellular antigen markers for neurons and astroglia. After 1–2 weeks *in vitro*, 95–100% of the cells in the reaggregates expressed characteristic features of granule neurons, with a small cell soma, 4–6 μm in diameter. By immunocytochemistry, virtually all of the cells expressed the neuronal antigen neural cell adhesion molecule (NCAM); a majority of the cells expressed L1 and TAG-1 (Gao *et al.*, 1991). Staining with antibodies against the glial filament protein indicated that the number of glial cells in the reaggregates remained at 1–2%, a value that corresponded to the average glial contamination at plating. This suggested that the immature granule cell population we purified gave rise to neuronal progeny and not to astroglial cells.

A distinguishing feature of the granule neuron is its utilization of the neurotransmitter glutamate rather than γ-aminobutyric acid (GABA), the transmitter system expressed by all other cerebellar neurons (Fonnum *et al.*, 1970; Kelly *et al.*, 1975; McLaughlin *et al.*, 1975; Hokfelt and Lungdahl, 1972). To examine whether proliferating granule neuroblasts gave rise to GABAergic neurons *in vitro*, we analyzed their expression of glutamate decarboxylase (GAD), the synthetic enzyme for GABA. Immunostaining of granule cells cultured in reaggregates for 1–2 weeks with antisera against GAD (Gottlieb *et al.*, 1986; Chang and Gottlieb, 1988) did not reveal the generation of GABAergic neurons *in vitro* (Gao *et al.*, 1991). As a positive control, we demonstrated immunostaining of Purkinje cells *in vitro* and of Purkinje cells, basket, stellate and Golgi II neurons in tissue sections at P5 and P12 with the antisera we used to stain cells in reaggregates *in vitro*.

INITIATION OF NEURONAL DIFFERENTIATION: THE NEUROLOGICAL MUTANT MOUSE *weaver*

To examine whether local interactions among progenitor cells in the EGL initiated neuronal differentiation, we measured neurite production of purified EGL cells in dispersed cell culture, in collagen gels, and in cellular reaggregates. As described previously, in the absence of growth factors, very little neurite extension was seen when EGL cells were dispersed on culture surfaces treated with PLYS or laminin. Similarly, after 24–48 hr *in vitro*, very little neurite extension occurred when the cells were plated in collagen, suggesting that diffusible factors were not sufficient to initiate neurite extension. By contrast, when cellular reaggregates of EGL cells were transferred to adhesive surfaces, a halo of neurites was evident around the perimeter of the reaggregate (Fig. 4A). This suggested that cell interactions among the precursor cells within the reaggregates induced neuronal differentiation (Gao *et al.*, 1991).

To determine whether the signal for granule cell neuronal differentiation was cell autonomous, we analyzed cells with phenotypic

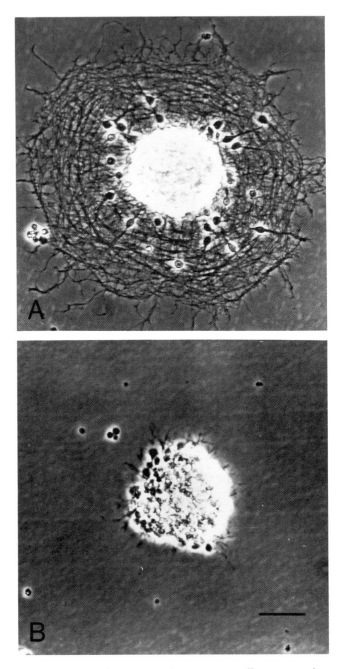

Fig. 4. Neurite outgrowth by *weaver* EGL precursor cells is impaired *in vitro*. EGL cells were purified from wild-type or *weaver* mouse cerebellum on P5 and maintained in cellular reaggregates for 48 hr, after which they were transferred to a substratum coated with polylysine to promote neurite extension. Thirty-six hours after transfer to the

defects in neuronal differentiation. At present, the neurological mutant mouse *weaver* provides the most opportune mutant for this purpose. In the *weaver* cerebellum, granule cell precursors proliferate
normally in the superficial layer of the EGL (Rezai and Yoon, 1972)
and move into the deepest layers of the EGL (Rakic and Sidman,
1973), but fail to extend neurites (Willinger and Margolis, 1985) or to
migrate away from the EGL (Rakic and Sidman, 1973; Sotelo and
Changeux, 1974; Hatten *et al.*, 1984b). Experimental analysis of the
site of action of the *weaver* gene, by mixing normal and mutant granule cells with normal and mutant astroglial cells *in vitro* (Hatten *et
al.*, 1986) and by the production of homozygous *weaver* chimeras
(Goldowitz and Mullen, 1982; Goldowitz, 1989), demonstrated that
the failure of *weaver* EGL neuronal precursor migration results from
defects in the differentiation of granule cell precursors, not from defects in neuron–glia interactions.

weaver EGL CELLS FAIL TO UNDERGO NEURONAL DIFFERENTIATION *IN VITRO*

To examine the role of cell–cell interactions in *weaver* EGL precursor cell neuronal differentiation, we purified EGL precursor cells
from the midline portion of the cerebellar cortex of P5–P7 B6CBA-
A^w-J-wv *(wv/wv)* animals by Percoll gradient sedimentation and panning techniques (Hatten, 1985, 1987; Hatten *et al.*, 1986) and cultured the cells in a reaggregate culture system (Gao *et al.*, 1991). As
seen *in vivo*, when *weaver* EGL cells were cultured in the reaggregate system, neurite extension failed (Fig. 4B).

weaver CELLS EXTEND NEURITES WHEN TRANSPLANTED INTO REAGGREGATES OF WILD-TYPE CELLS

To determine whether wild-type EGL precursor cells provide a
signal that induces neurite extension, we mixed *weaver* granule cell
precursors with wild-type precursor cells at a ratio of 1:10 and examined neurite production by *weaver* cells. To visualize *weaver* neurite

polylysine substratum, whereas wild-type EGL cells (A) extend a dense halo of neurites
onto the culture surface, *weaver* EGL cells (B) fail to extend neurites. Phase contrast
microscopy. Bar: 50 μm. Reprinted with permission from Gao *et al.* (1992).

production, we labeled *weaver* granule cell precursors with PKH26-GL (Zynaxis Cell Science, Inc., Malvern, PA), a stable membrane-soluble dye (Horan and Slezak, 1989). In reaggregates of wild-type cells, *weaver* cells extended neurites that reached the perimeter of the skirt of wild-type neurites bordering the cellular reaggregate (Gao *et al.*, 1992). This suggested that cell–cell interactions with wild-type EGL cells rescued the phenotypic defect in *weaver* neurite extension.

To confirm the results of dye-labeling experiments, we carried out interspecies mixing experiments. In these experiments, mouse *weaver* EGL cells were mixed with rat wild-type EGL cells at a ratio of 1:10 and mouse *weaver* cell neurites were identified by immuno-staining with a mouse-specific monoclonal antibody M6 (Lund *et al.*, 1985; Baird *et al.*, 1992). After mixing with rat EGL cells, *weaver* mouse cells extended neurites comparable in length to wild-type cells (Gao *et al.*, 1992). Thus both dye-labeling and interspecies-marking methods provided evidence that wild-type EGL precursors rescued the *weaver* phenotypic defect in neurite production *in vitro*, suggesting that the *weaver* gene acts non-autonomously

MEMBRANES FROM WILD-TYPE CELLS RESCUE THE *weaver* DEFECT IN NEURITE OUTGROWTH

To examine whether the induction of *weaver* cell neurite out-growth was by a signal bound to the membrane of wild-type cells, we mixed *weaver* EGL cells with membranes of wild-type cells or *weaver* cells, reaggregated the cells on an untreated culture surface for 48 hr, transferred the reaggregates to a PLYS substratum, and examined the extent of neurite outgrowth. The addition of membrane material purified from wild-type, but not mutant EGL, cells induced neurite outgrowth in a dose-dependent manner (Fig. 5). The rescue of *weaver* neurite extension by membrane material from wild-type cells suggested that the induction of neuronal differentiation was by a membrane-bound signal.

weaver CELLS MIGRATE ALONG GLIAL FASCICLES AFTER TRANSPLANTATION INTO REAGGREGATES OF WILD-TYPE CELLS

To examine whether mixing *weaver neuronal* precursor cells with wild-type *neuronal* precursor cells would rescue the phenotypic de-

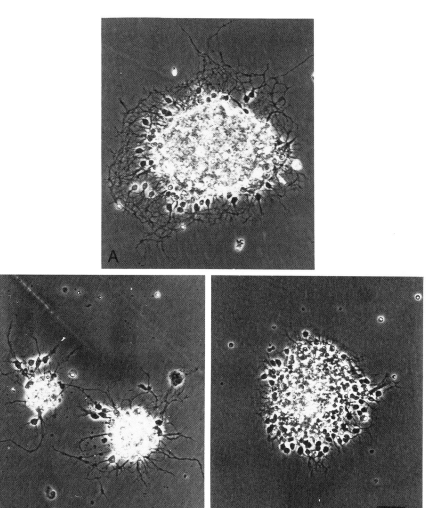

Fig. 5. Membranes from wild-type EGL cells rescue neurite outgrowth by reaggregated *weaver* cells. EGL cells were purified from *weaver* mouse cerebellum on P5 and plated in medium containing membranes from wild-type EGL cells (A,B) or medium conditioned by wild-type cells (C). After 48 hr, the reaggregates were transferred to a polylysine-coated substratum to promote neurite extension. Thirty-six hours after transfer to the polylysine surface, the cultures were fixed with glutaraldehyde (2%) and neurite outgrowth was examined by microscopy. In the presence of 200 μg/ml of wild-type membrane material (A), *weaver* EGL cells show robust neurite extension. After the addition of 100 μg/ml of wild-type membrane material (B), fewer *weaver* cells are induced to extend neurites. The addition of medium conditioned by wild-type EGL cells fails to induce *weaver* cell neurite extension (C). Phase-contrast microscopy. Bar: 30 μm. Reprinted with permission from Gao *et al.* (1992).

fect in cell migration, we labeled *weaver* EGL cells with PKH26-GL, mixed the mutant cells with wild-type EGL cells, and added wild-type astroglial cells to provide a substrate for migration (Hatten and Sidman, 1978; Trenkner and Sidman, 1977). After transplantation into cellular reaggregates of wild-type cells, labeled, mutant cells migrated out onto the glial fascicles interconnecting the reaggregates (Fig. 6). Visualization of dye-labeled, *weaver* cells by fluorescence microscopy showed that labeled, mutant cells expressed the cytological features of migrating granule cells (see below), including the formation of a close apposition with the glial fiber and the extension of a leading process.

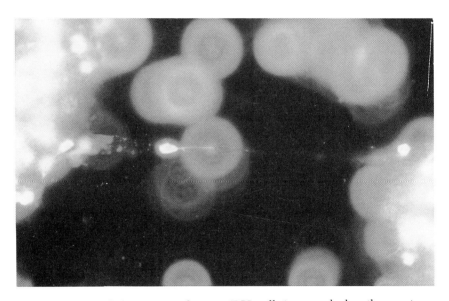

Fig. 6. Glia-guided migration of *weaver* EGL cells is rescued when they are transplanted into cellular reaggregates of wild-type EGL cells. EGL cells were purified from *weaver* mouse cerebellum on P5, labeled with PKH26-GL, recombined at a 1:10 ratio with a mixture of wild-type EGL cells and wild-type astroglial cells (10 EGL cells/glial cell), and cultured on an untreated culture surface for 36 hr. Under these conditions, cellular reaggregates are interconnected by fascicles composed of glial fibers, and granule neurons migrate from the cellular reaggregates out onto the glial fibers (Trenkner and Sidman, 1977). By fluorescence microscopy a labeled *weaver* neuron is seen migrating along the fascicle. The dye-labeled *weaver* cell displays the profile characteristic of migrating neurons (Hatten, 1990), apposing the neural soma against the glial fiber and extending a leading process in the direction of migration (toward the bottom of the field as shown). Reprinted with permission from Gao *et al.* (1992).

EXPRESSION OF TAG-1 AND ASTROTACTIN INCREASES WHEN *weaver* CELLS ARE COCULTURED WITH WILD-TYPE CELLS

The rescue of *weaver* cell neurite extension and glia-guided migration observed when the cells were mixed with wild-type precursor cells raised the question as to whether rescued mutant cells expressed markers of later stages of granule cell neuronal differentiation. To address this question, we immunostained dye-labeled mutant cells with antibodies against the transient axonal glycoprotein (TAG-1), an axonal glycoprotein of the IgG superfamily that promotes neurite outgrowth (Furley *et al.*, 1990), and astrotactin, a neuron–glia ligand that functions in glial-guided neuronal migration *in vitro* (Fishell and Hatten, 1991). By double-labeling, antibodies against both TAG-1 and astrotactin strongly labeled rescued *weaver* neurites. As the high density of wild-type neurites made it difficult to visualize double-labeled, *weaver* cell neurites reliably, we induced mutant cell neurite outgrowth with wild-type cell membranes and examined TAG-1 expression by rescued mutant cells. To preclude staining of wild-type membranes with anti-TAG-1 antibodies, we released TAG-1 from wild-type membranes by treatment with phospholipase C (PL-C) and induced mutant cell neurite extension with TAG-1-negative membranes. By immunostaining, after induction with wild-type membranes, all of neurites extended by *weaver* cells were strongly labeled with TAG-1 (Gao *et al.*, 1992).

The rescue of *weaver* cell neurite production and migration by coculture with wild-type cells indicates that action of the *weaver* gene is not cell autonomous, suggesting that the *weaver* gene encodes a signal for EGL neuronal differentiation or a gene that modulates the expression of such a factor. The requirement for close apposition of mutant cells with wild-type cells suggests that contact interactions among EGL precursor cells are involved in the initial steps of neuronal differentiation. This interpretation is consistent with the fact that initiation of neuronal differentiation occurs in compact ventricular zones in the developing mammalian brain, zones where close cell appositions can be maintained (His, 1889; Ramon y Cajal, 1911; Sauer, 1935; Boulder Committee, 1970).

The finding that cell–cell interactions with *neuronal* precursors induced *weaver* neuronal migration presents a new interpretation of the influence of the *weaver* locus on neuronal migration. As previous *in vitro* mixing experiments showed that wild-type astroglial cells fail to rescue the *weaver* defect in migration (Hatten *et al.*, 1986), the finding that cell–cell interactions with neuronal precursors rescue the migration of *weaver* cells along astroglial fibers (Gao *et al.*, 1992) suggests that cell–cell interactions among EGL neuronal precursors induce the expression of genes required for cell migration.

FEATURES OF MIGRATING GRANULE NEURONS

To examine the control of precursor cell migration from the EGL along the fibers of the Bergmann glial cells, we prepared cultures of purified, immature granule cells and cerebellar glia (Edmondson and Hatten, 1987). A key feature of our *in vitro* system is the small volume (20–50 μl) in which we culture the cells. In these microcultures, astroglial cells differentiate into highly elongated, radial forms which resemble the glial forms shown to support neuronal migration in brain (Rakic, 1971). These studies provided evidence that astroglial fibers provide a scaffold for neuronal migration. By video microscopy, the migration of living, identified neurons along single glial fibers could be observed in real time (Hatten *et al.*, 1984b; Edmondson and Hatten, 1987). With video-enhanced differential interference contrast (AVEC-DIC) microscopy, migrating neurons are seen to express a highly extended bipolar shape along the glial fiber, forming a close apposition with the glial process along the length of the neuronal cell soma and extending a leading process in the direction of migration (Gregory *et al.*, 1988; Edmondson and Hatten, 1987; Gasser and Hatten, 1990; Hatten, 1990). The leading process is a highly motile structure, extending and retracting short filopodia and lamellopodia that enfold the glial fiber. Further, organelles flow from the cell body forward into the leading process. Thus, the migrating neuron appears to move along the glial fiber by forming a "footpad" beneath the cell body and extending a tapered, leading process along the glial guide (Gregory *et al.*, 1988).

Although the leading process resembles the growth cone of a neurite in its motility, our *in vitro* studies suggest that it differs in several aspects. First, whereas the growth cone is an expansive ending of a thin neurite, the leading process is contoured to the dimensions of the glial fiber. Second, whereas the growth cone is the site of growth and extension, the leading process is simply a tapered rostral portion of the cell. Third, whereas the growth cone is the site of adhesion of the growing neurite, the cell body is the site of adhesion of the migrating neuron. A common feature of the growth cone and the leading process is that both guide the directionality of the neuron (Edmondson and Hatten, 1987).

MOLECULAR COMPONENTS OF GLIA-GUIDED NEURONAL MIGRATION

To define neuron–glia adhesion systems in CNS glial-guided neuronal migration, we used our *in vitro* assays to identify an immune activity, which we named anti-astrotactin, that blocks neuronal migration *in vitro* (Edmondson *et al.*, 1988; Fishell and Hatten, 1991). Western blot and immunoprecipitation analyses indicate that the blocking antibodies recognize a neuronal glycoprotein with an apparent molecular mass of approximately 100 kDa (Edmondson *et al.*, 1988). Antiastrotactin antibodies block the binding of neuronal membranes to glial cells (Stitt and Hatten, 1990) and the establishment of neuron–glia contacts *in vitro* (Edmondson *et al.*, 1988). To examine the contribution of astrotactin and neural cell adhesion systems to neuronal migration along glial fibers, we developed an assay to quantify the migration of large populations of granule cells and examined the perturbation of neuronal migration by antibodies against cell adhesion molecules (Fishell and Hatten, 1991).

Three general classes of receptor systems were analyzed in antibody perturbation assays, the neuron–glial adhesion ligand astrotactin; the neural cell adhesion molecules of the IgG superfamily NCAM, L1, and TAG-1; and the β_1 subunit of the integrin family (Fishell and Hatten, 1991). In the absence of immune activities, migrating cerebellar granule neurons had an average *in vitro* migration rate of 12 μm/hr with individual neurons exhibiting migration rates over a range between 0 to 70 μm/hr. The addition of antiastrotactin antibodies (or Fabs) significantly reduced the mean rate of neuronal migration by 61%, resulting in 80% of the neurons having migration rates below 8 μm/hr (Fig. 7A). By contrast, blocking antibodies (or Fabs) against L1, NCAM, TAG-1, or β_1-integrin, individually or in combination, did not reduce the rate of neuronal migration (Fig. 7B).

By video-enhanced contrast differential interference contrast microscopy the effects of antiastrotactin antibodies were rapid. Within 15 min of antibody application, streaming of cytoplasmic organelles into the leading process arrested, the nucleus shifted from a caudal to a central position, and the extension of filopodia and lamellopodia along the leading process ceased (Fig. 8). Correlated video and electron microscopy suggested that the mechanism of arrest by antiastrotactin antibodies involved the failure to form new adhesion sites along the leading process and the disorganization of cytoskeletal components. These results suggest that astrotactin acts as a neuronal receptor for granule neuron migration along astroglial fibers (Fishell and Hatten, 1991).

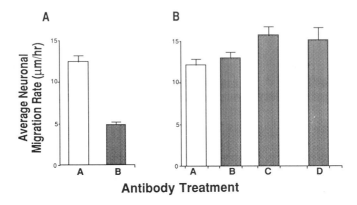

Fig. 7. Antiastrotactin antibodies reduce the rate of granule neuron migration. In (A) the percent distribution of migratory rates of a population of granule neurons *in vitro* is shown. Cerebellar granule neurons migrate at speeds ranging from 0 to 70 μm/hr. For this analysis, populations of cells enriched for cerebellar granule neurons were plated on Matrigel-coated surfaces. Thirty to forty hours after plating, the rate of neuronal migration of large numbers of granule neurons was tracked using phase-contrast microscopy, and the speed and frequency of migration of populations of neurons was tabulated. In (B) the mean rate of migration of granule neurons in the presence of L1 antibodies (treatment B); NCAM antibodies (treatment C) (both at antibody concentrations of 1.0 mg/ml); and a combination of L1, NCAM, and β_1 integrin antibodies (treatment D) (each antibody at a concentration of 0.5 mg/ml) is compared to the mean rate of migration under control conditions (treatment A). None of these antibody treatments significantly change the rate of granule neuron migration. Reprinted with permission from Fishell and Hatten (1991).

SPATIOTEMPORAL PATTERN OF EXPRESSION OF ASTROTACTIN

By immunocytochemical localization studies, the pattern of expression of astrotactin differs from that of neural adhesion systems shown to promote axon outgrowth, as it is expressed only by cells undergoing glial-guided neuronal migration and assembly (Fishell *et al.*, 1992). Moreover in the cerebellum, a specific spatial distribution of astrotactin and neural cell adhesion molecules is seen for the granule neuron. Whereas the neural cell adhesion molecules are expressed on the axons of the granule cell (Persohn and Schachner, 1987), astrotactin expression is restricted to the soma and leading process of migrating granule cells. This spatial distribution is consistent with functional studies that showed that antibodies against astrotactin, but

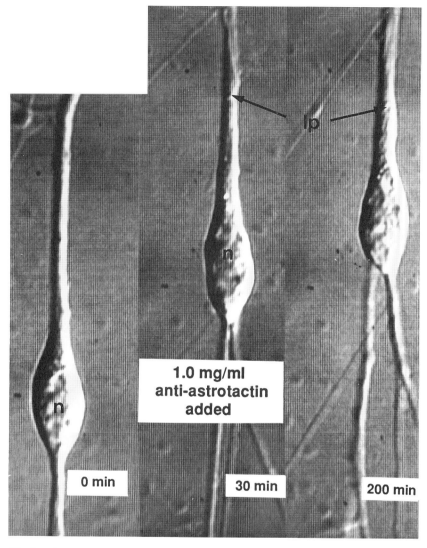

Fig. 8. Antiastrotactin antibodies rapidly arrests neuronal migration *in vitro*. A single migrating granule neuron is shown before, during, and after the addition of 1.0 mg/ml of antiastrotactin antibodies, using high-power DIC optics (100×/1.6). Prior to the addition of antibodies, this neuron displays the cytology and activity of a typically migrating neuron. The addition of antiastrotactin antibodies (at 30 min) rapidly arrested further neuronal migration. Three hours after the addition of the antibodies, the leading process (lp) of the migrating neuron collapsed and started to withdraw. Similarly the neuronal nucleus (n) moved from its typical caudal position, to a more central position in the cell soma. Reprinted with permission from Fishell and Hatten (1991).

not neural cell adhesion molecules, block neuron–glia binding and neuronal migration (Fishell and Hatten, 1991).

The specific spatiotemporal expression of the axonal glycoprotein TAG-1 and astrotactin by granule cells initiating axon extension and glial-guided migration in the EGL suggests that discrete stages of neuronal differentiation are marked by the expression of molecular components required for axon guidance and migration. To further define the program of gene expression leading to granule neuron differentiation in the developing cerebellum, we carried out molecular biological analysis to provide molecular markers in specific developmental stages.

CHANGING PATTERNS OF GENE EXPRESSION DEFINE FOUR STAGES OF GRANULE NEURON DIFFERENTIATION

To identify markers for granule cells in different stages of neuronal development, we constructed cDNA expression libraries from immature granule cells, purified from mouse cerebellum on postnatal days 3–5 (Hatten, 1985), and used novel antibody screening methods to identify cDNA clones that mark specific stages of granule neuron differentiation (Kuhar *et al.*, 1992). In the initial screen of 52 purified phage clones, 39 clones were found to be unique by restriction mapping of polymerase chain reaction (PCR)-amplified inserts. Unique clones were subcloned and nucleotide sequences were determined using double-stranded dideoxy sequencing. Approximately 250 nucleotides of sequence were obtained from both the 5′ and 3′ ends of each clone. A search of all current GenBank nucleotide sequences indicated that 28 of the 39 clones were novel genes (Kuhar *et al.*, 1992).

Northern blot analysis of poly(A)$^+$ RNA purified from various tissues, brain regions, and developmental stages revealed that of 34 clones analyzed, all were expressed at high levels in cerebellum. Four of the clones were expressed at higher levels in the cerebellum and forebrain; 3 clones were detectable only in cerebellum. Many clones were expressed in other tissues as well, suggesting that the screen generated cDNAs expressed at high levels in cerebellum, but were not necessarily unique to cerebellum (Kuhar *et al.*, 1993).

In situ analysis of the expression of developmentally regulated clones in cerebellar tissue at P8 offered the opportunity to visualize individual cells in each of the stages of granule neuron development.

As shown in Fig. 9, a restricted pattern of expression was observed for seven clones we isolated, suggesting that different cDNAs were expressed at different stages of granule neuron differentiation. Four discrete stages of differentiation were revealed by the cDNAs analyzed. The first stage of granule cell differentiation (Fig. 9A), proliferation of progenitor cells in the outer portion of the EGL, was represented by cDNAs that labeled a thin layer of cells along the pial surface, including GC 10, 27, 60, and 61 (Fig. 9A). This suggests that the four novel clones that labeled cells along the cerebellar surface were expressed in proliferating granule cell precursors.

Although some clones were expressed throughout the EGL, including GC-1, other clones were localized to a zone just beneath the superficial portion of the EGL, a zone where young granule cells express late neuronal markers, including the axonal glycoprotein

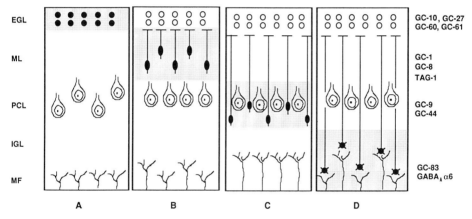

Fig. 9. Changing patterns of gene expression define four stages of granule cell differentiation. *In situ* hybridization analysis of the pattern of expression of cDNAs expressed in granule cells during develement indicates a discrete pattern of localization for four classes of cDNAs. At P7, (A) GC10, GC27, GC60, and GC61 mRNA are localized to proliferating granule cells in the superficial layer of the EGL (solid circles in shaded zone). (B) GC-1, GC-8, and TAG1 mRNA are localized to postmitotic and migrating granule cells at the internal surface of the EGL and in the molecular layer, respectively (solid circles in shaded zone). (C) GC9 and GC44 mRNA are both localized to a previously undescribed transient layer of granule cells (solid circles in shaded zone) terminating migration just below the Purkinje cell bodies and above the GC-83 and g1 subunit GABA$_A$ receptor-positive cells in the IGL (D). *In situ* hybridizations were performed as described for the Genius system (Boehringer-Mannheim) and processed for colorimetric detection using NBT and X-phosphate (Kuhar *et al.*, 1992). Ingrowing afferent axons, the mossy fibers, are shown at the bottom of each panel.

TAG-1 (Furley *et al.*, 1990), the neuron–glia ligand astrotactin (Edmondson *et al.*, 1988), and the intermediate filament protein α-internexin (Kaplan *et al.*, 1990), and begin to extend parallel fibers. A third group of cDNAs, including GC9 and GC44, showed patterns of expression that were restricted to cells in the upper portion of the internal granular layer (IGL) (Fig. 9C) that is contiguous with the overlying Purkinje cell layer. As shown in Fig. 9, whereas clones GC9 and 44 labeled cells in the upper aspect of the IGL, expression of the GABA$_A$ α$_6$ receptor subunit mRNA was restricted to cells that were located beneath cells labeled with GC9 and GC44 (Fig. 9D, Kuhar *et al.*, 1992). The restricted pattern of expression of GC9 and GC44 and the GABA$_A$ α$_6$ subunit permitted a dissection of the IGL into two layers, an upper layer, IGLa, and a lower layer, IGLb, where the most differentiated cells are located, and suggested that genes that were up-regulated during cerebellar development marked a fourth stage of granule neuron differentiation.

The program of gene expression in cerebellar granule cells revealed in this analysis demonstrates that there are at least four distinct stages of granule cell differentiation. Analysis of the regulation of these novel cDNAs in neurological mutant mice and *in vitro* functional assay systems using purified cerebellar cell populations (Hatten, 1985; Gao *et al.*, 1991; Fishell and Hatten, 1991; Baird *et al.*, 1992) should allow the identification of the molecular signals for granule neuron differentiation. In particular, they should allow the identification of membrane-associated signals that regulate precursor cell proliferation and induce neuronal differentiation (Gao *et al.*, 1991, 1992) and of transcription factors that regulate the expression genes at specific stages of granule neuron differentiation.

ACKNOWLEDGMENTS

Supported by NIH Grant NS 15429 (M.E.H.), the PEW Neuroscience Program (M.E.H.), and the Howard Hughes Medical Institute (N.H.)

REFERENCES

Altman, J. (1972). Postnatal development of the cerebellar cortex in the rat. I. The external germinal layer and the transitional molecular layer. *J. Comp. Neurol.* **145**, 353–514.

Anderson, D. J. (1989]. The neural crest lineage problem: Neuropoiesis? *Neuron* **3**, 1–12.

Angevine, J. B., Jr., and Sidman, R. L. (1961). Autoradiographic study of cell migration during histogenesis of cerebral cortex in the mouse. *Nature* **192**, 766–768.

Baird, D., Hatten, M. E., and Mason, C. A. (1992). Cerebellar target neurons provide a stop-signal for afferent neurite extension in vitro. *J. Neurosci.* **12**, 619–634.

Boulder Committee (1970). Embryonic vertebrate central nervous system: Revised terminology. *Anat. Rec.* **166**, 257–262.

Chang, Y.-C., and Gottlieb, D. I. (1988). Characterization of the proteins purified with monoclonal antibodies to glutamic acid decarboxylase. *J. Neurosci.* **8**, 2123–2130.

Edmondson, J. C., and Hatten, M. E. (1987). Glial-guided granule neuron migration *in vitro:* A high-resolution time-lapse video microscopic study. *J. Neurosci.* **7**, 1928–1934.

Edmondson, J. C., Liem, R. K. H., Kuster, J. E., and Hatten, M. E. (1988). Astrotactin: A novel neuronal cell surface antigen that mediates neuron-astroglial interactions in cerebellar microcultures. *J. Cell Biol.* **106**, 505–517.

Fishell, G., and Hatten, M. E. (1991). Astrotactin provides a receptor system for glia-guided neuronal migration. *Development* **113**, 755–765.

Fonnum, F., Storm-Mathisen, J., and Walberg, F. (1970). Glutamate decarboxylase in inhibitory neurons: A study of the enzyme in Purkinje cell axons and boutons in the cat. *Brain Res.* **20**, 59–275.

Fujita, S. (1967). Quantitative analysis of cell proliferation and differentiation in the cortex of the postntal mouse cerebellum. *J. Cell Biol.* **32**, 277–287.

Fujita, S., Shimada, M., and Nakanuna, T. (1966). [^{3}H]Thymidine autoradiographic studies on the cell proliferation and differentiation in the external and internal granular layers of the mouse cerebellum. *J. Comp. Neurol.* **128**, 191–209.

Furley, A. J., Morton, S. B., Manalo, D., Karagogeos, Dodd, J., and Jessell, T. M. (1990). The axonal glycoprotein TAG-1 is an immunoglobulin superfamily member with neurite outgrowth promoting activity. *Cell* **61**, 157–170.

Gao, W.-Q., Heintz, N., and Hatten, M. E. (1991). Cerebellar granule cell neurogenesis is regulated by cell-cell interactions in vitro. *Neuron* **6**, 705–715.

Gao, W.-Q., Liu, X.-L., and Hatten, M. E. (1992). The *weaver* gene encodes a non-autonomous signal for CNS neuronal differentiation. *Cell* **68**, 841–854.

Gasser, U. E., and Hatten, M. E. (1990). CNS neurons migrate on astroglial fibers from heterotypic brain regions *in vitro. Proc. Natl. Acad. Sci. U.S.A.* **87**, 4543–4547.

Goldowitz, D. (1989). The *weaver* granuloprival phenotype is due to intrinsic actin of the mutant locus in granule cells: Evidence from homozygous *weaver* chimeras. *Neuron* **2**, 1565–1575.

Goldowitz, D., and Mullen, R. J. (1982). Granule cell as a site of gene action in the *weaver* mouse cerebellum: Evidence from heterozygous mutant chimeras. *J. Neurosci.* **2**, 1474–1485.

Gottlieb, D. I., Chang, Y.-C., and Schwob, J. E. (1986). Monclonal antibodies to glutamic acid decarboxylase *Proc. Natl. Acad. Sci. U.S.A* **83**, 8808–8812.

Gratzner, H. G. (1982). Monoclonal antibody to 5-bromo and 5-iodiodeoxyuridine: A new reagent for detection of DNA replication. *Science* **218**, 474–475.

Gregory, W. A., Edmondson, J. C., Hatten, M. E., and Mason, C. A. (1988). Cytology and neuron-glial apposition of migrating cerebellar granule cells in vitro. *J. Neurosci.* **8**, 1728–1738.

Hatten, M. E. (1985). Neuronal regulation of astroglial morphology and proliferation *in vitro. J. Cell Biol.* **100**, 384–396.

Hatten, M. E. (1987). Neuronal regulation of astroglial proliferation is membrane-mediated. *J. Cell Biol.* **104**, 1353–1360.

Hatten, M. E. (1990). Riding the glial monorail: A common mechanism for glial-guided migration in different regions of the developing brain. *Trends Neurosci.* **13**, 179–184.

Hatten, M. E., and Sidman, R. L. (1978). Cell reassociation behavior and lectin-induced agglutination of embryonic mouse cells in different brain regions. *Exp. Cell Res.* **113**, 111–125.

Hatten, M. E., Liem, R. K. H., and Mason, C. A. (1984a). Two forms of cerebellar glial cells interact differently with neurons *in vitro. J. Cell Biol.* **98**, 193–204.

Hatten, M. E., Liem, R. K. H., and Mason, C. A. (1984b). Defects in specific associations between astroglia and neurons occur in microcultures of *weaver* mouse cerebellar cells. *J. Neurosci.* **4**, 1163–1172.

Hatten, M. E., Liem, R. K. H., and Mason, C. A. (1986). *Weaver* mouse cerebellar granule neurons fail to migrate on wild-type astroglial processes *in vitro. J. Neurosci.* **6**, 2676–2683.

His, W. (1889). Die neuroblasten und deren entstehung im embryonalen mark. *Abh. Math. Phys. Kl. Konigl. Saechs Ges. Wiss.* **26**, 313–372.

Hokfelt, T., and Lungdahl, A. (1972). Autoradiographic identification of cerebral and cerebellar cortical neurons accumulating labeled gamma amino-butyric acid (3H-GABA) *Exp. Brain Res.* **14**, 354–362.

Horan, P. K., and Slezak, S. E. (1989). Stable cell membrane labelling. *Nature* **340**, 167–168.

Kaplan, M. P., Chin, S. S. M., Fliegner, K. H., and Liem, R. K. H. (1990). α-internexin, a novel neuronal intermediate filament protein, precedes the low molecular weight neurofilament protein (NF-L) in the developing rat brain. *J. Neurosci.* **10**, 2735–2748.

Kelly, J. S., Dick, F., and Schon, F. (1975). The autoradiographic localization of the GABA-releasing nerve terminals in cerebellar glomeruli. *Brain Res.* **85**, 255–259.

Kuhar, S. G., Feng, L., Vidan, S., Ross, E. M., Hatten, M. E., and N. Heintz. (1993). Developmentally regulated cDNAs define four stages of cerebellar granule neuron differentiation. *Development* **117**, 97–104.

LeDouarin, N. M. (1982). "The Neural Crest." Cambridge Univ. Press, Cambridge.

Lund, R. D., Chang, F-LF, Hankin, M., and Lagenaur, C. F. (1985). Use of a species-specific antibody for demonstrating mouse neurons transplanted to rat brains. *Neurosci. Lett.* **61**, 221–226.

McLaughlin, B. J., Wood, J. G., Saito, K., Roberts, E., and Wu, Y. Y. (1975). The fine structure of glutamate decarboxylase in developing axonal processes and pre-synaptic terminals of rodent cerebellum. *Brain Res.* **85**, 355–371.

Miale, I., and Sidman, R. L. (1961). An autoradiographic analysis of histogenesis in the mouse cerebellum. *Exp. Neurol.* **4**, 277–296.

Patterson, P. (1978). Environmental determination of autonomic neurotransmitter functions. *Annu. Rev. Neurosci.* **1**, 1–17.

Persohn, E., and Schachner, M. (1987). Immunoelectron microscopic localization of the neural cell adhesion molecules L1 and NCAM during postnatal development of the cerebellum. *J. Cell Biol.* **105**, 569–576.

Rakic, P. (1971). Neuron-glia relationship during granule cell migration in developing cerebellar cortex: A Golgi and electron microscopic study in Macacus rhesus. *J. Comp. Neurol.* **141**, 283–312.

Rakic, P., and Sidman, R. L. (1973). Sequence of developmental abnormalities leading to granule cell deficit in cerebellar cortex of *weaver* mutant mice. *J. Comp. Neurol.* **152,** 103–132.

Ramon y Cajal, S. (1889). Sobre las fibras nerviosas de la capa granulosa del cerebelo. *Rev. Trim. Histol. Norm. Pathol.* Nos. 3 and 4.

Ramon y Cajal, S. (1911). "Histologie du Systeme Nerveux de l'Homme et des Vertebres," Maloine, Paris (reprinted by Consejo Superior de Investigaciones Cientificas, Madrid, 1955).

Rezai, Z., and Yoon, C. H. (1972). Abnormal rate of granule cell migration in cerebellum of *"weaver"* mutant mice. *Dev. Biol.* **29,** 17–26.

Sauer, F. C. (1935). The cellular structure of the neural tube. *J. Comp. Neurol.* **63,** 12–23.

Sidman, R. L., and Rakic, P. (1973). Neuronal migration with special reference to human brain: A review. *Brain Res.* **62,** 1–35.

Sotelo, C., and Changeux, P. (1974). Bergmann fibers and granule cell migration in the cerebellum of homozygous *weaver* mutant mouse. *Brain Res.* **77,** 484–494.

Stitt, T. N., and Hatten, M. E. (1990). Antibodies that recognize astrotactin block granule neuron binding to astroglia. *Neuron* **3,** 639–639.

Trenkner, E., and Sidman, R. (1977). Histogenesis of mouse cerebellum in microwell culture: Cell reaggregation and migration, fiber and synapse formation. *J. Cell Biol.* **75,** 915–940.

Willinger, M., and Margolis, D. M. (1985). Effect of the *weaver* (wv) mutation on cerebellar neuron differentiation. I. Quantitative observations of neuron behavior in culture. *Dev. Biol.* **107,** 156–172.

3

Mesoderm Induction and Pattern Formation in the Amphibian Embryo

IGOR B. DAWID, MICHAEL R. REBAGLIATI, AND
MASANORI TAIRA
Laboratory of Molecular Genetics
National Institute of Child Health and Human Development
National Institutes of Health
Bethesda, Maryland 20892

MESODERM INDUCTION AND THE ESTABLISHMENT OF THE BASIC BODY PLAN IN *Xenopus*

The establishment of pattern during the development of all animals, and especially of vertebrates, depends on cell interactions in which the fate of some cells is profoundly affected by signals from other cells. Such interactions are referred to as inductive events. Embryonic induction has been discovered in amphibians, first in the development of the eye, and most dramatically in the organizer experiments of Spemann and Mangold (1924) in which the dorsal blastopore lip of one gastrula embryo was transplanted into the ventral side of a host gastrula, resulting in the induction of a ssecondary embryonic axis. Because of its inducing properties the dorsal lip mesoderm is often referred to as the Spemann organizer. A major step in advancing our understanding induction and pattern formation in this system was achieved by the realization that the earliest inductive interaction in the amphibian embryo (and probably all vertebrate embryos) is required for the formation of the mesoderm (Fig. 1; Nieuwkoop, 1969, 1973; Sudarwati and Nieuwkoop, 1971; for recent reviews see Smith, 1989; Dawid *et al.*, 1990; Melton, 1991). This inductive event (or perhaps, series of events) also differentiates dorsal from ventral mesoderm, thus setting up one of the major axes of the embryonic body plan. Formation

33

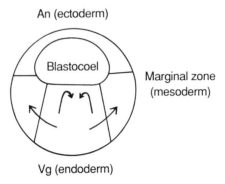

Fig. 1. A model showing that induction is required for mesoderm specification. Signals derived from the vegetal (Vg) region impinge on the marginal zone, specifying it as future mesoderm. The vegetal region itself will form primarily endoderm. The animal (An) region is not normally exposed to these signals since they cannot cross the blastocoel (bent arrows); animal cells thus follow their inherent fate to differentiate into ectoderm.

of the nervous system is initiated somewhat later during gastrula stages, when dorsal ectoderm is specified, most likely as a result of signals emanating from the mesoderm, to produce the neural plate (Gimlich and Cooke, 1983; Smith and Slack, 1983; Jacobson, 1984). During these stages the anteroposterior axis of the embryo is established through the migration of the dorsal mesoderm and its interaction with the overlying ectoderm. Thus, definition of the two major axes and elaboration of the basic body plan of the embryo require two interconnected though separable inductive processes in which the specification of dorsal and anterior structures is intimately associated.

The experimental system that has allowed the analysis of mesoderm induction is based on the work of Nieuwkoop (1969, 1973) and is illustrated in Fig. 2. The animal region of a blastula is fated to develop into ectoderm; if it is explanted and cultured in control medium it will follow its inherent fate and differentiate into a somewhat atypical epidermis. Culture of such animal explants ("animal caps") in contact with vegetal tissue leads to the production of all types of mesodermal derivatives. From this and similar experiments Nieuwkoop and others concluded that the vegetal pole of the embryo is the source of a signal or set of signals that respecifies the equatorial region (called the marginal zone) toward a mesodermal fate; the animal region is competent to respond to this signal but does not receive it during normal embryogenesis. These conclusions have been corroborated by subsequent

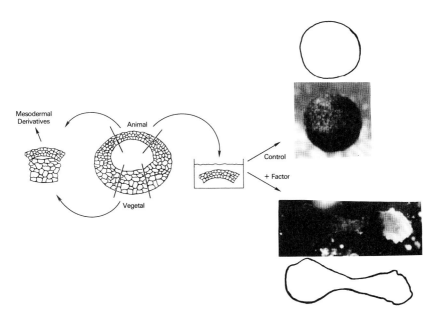

Fig. 2. The animal cap assay. The left-hand side shows the original Nieuwkoop experiment in which animal tissue from a midblastula embryo is cultured in association with vegetal tissue, giving rise to mesoderm. On the right-hand side the assay for soluble induction factors is illustrated. Animal caps in control media without inducing factor turn into spherical bodies that are composed of atypical epidermis. Induction factors lead to elongation and differentiation of mesodermal tissues. See text for further discussion.

work, including experiments with dissociated cells that showed a requirement for cell interaction in the specification of muscle (Sargent *et al.*, 1986).

Rapid progress has been made in analyzing the mechanism of mesoderm induction. This progress owes much to the discovery by Smith (1987) of the inducing ability of the culture medium of XTC cells. Shortly after this observation, fibroblast growth factor (FGF) was found to have mesoderm-inducing ability, although this factor induces only ventral and lateral mesoderm and thus cannot by itself account for the entire inducing signal (Slack *et al.*, 1987; Kimelman and Kirschner, 1987). Members of the transforming growth factor β (TGF-β) family were subsequently implicated as strong mesoderm inducers; in particular, TGF-β_2 is capable of inducing all types of mesoderm in the animal cap assay (Rosa *et al.*, 1988). More recently, the distant relative of TGF-β, activin A, was identified as the most potent mesoderm inducer (Asashima *et al.*, 1990) and shown to represent the active principle in XTC medium (Smith *et al.*, 1990). Activin B mRNA is expressed only in the late blastula and activin A mRNA in the gastrula (Thomsen *et al.*, 1990), apparently too late for the initial stages of mesoderm induction, but activin-like protein with inducing ability has been found in unfertilized eggs and early embryos (Asashima *et al.*, 1990). Thus one or more forms of activin are likely to participate in mesoderm induction *in vivo*. Additional members of the TGF-β family are expressed in the embryo (Rebagliati *et al.*, 1985; Weeks and Melton, 1987; Köster *et al.*, 1991), but their role in mesoderm formation is not clear at this time.

Recently it was shown that mRNAs encoding members of the *Wnt* family of proteins have dorsal axis-inducing properties when injected into the ventral side of early frog embryos (Sokol *et al.*, 1991; Smith and Harland, 1991). Two types of RNA were effective in this assay: mammalian *Wnt-1* RNA and *Xenopus Wnt-8* RNA. The *Wnt-8* gene is expressed during normal embryogenesis in the ventral marginal zone of blastula and gastrula embryos, and it is therefore unclear why the injection of more of the same RNA at the location where it normally occurs should induce dorsal differentiation. Thus, the role of *Wnt* family proteins in normal embryogenesis is not clear, but the striking phenotypes that are caused by *Wnt* RNA injection suggest that one or more members of this gene family are involved in mesoderm induction and dorsal axis specification.

MESODERM INDUCTION AND THE FORMATION OF THE DORSOVENTRAL AND ANTEROPOSTERIOR AXES

During the first day of development the frog embryo not only differentiates a variety of cell types and tissues but also establishes its major axes, i.e., its body plan. How this happens, in the frog or any animal, is a key question of developmental biology. It is clear that axis formation is intimately linked to mesoderm induction and differentiation, as discussed above. The specific question addressed in this section deals with the morphological consequences of the application of inducing factors to competent tissue: Does activin simply induce a disorganized mass of mesodermal derivatives or does it elicit the arrangement of these tissues along embryonic axes? This question was brought into focus by a report by Sokol *et al.* (1990), showing that a crude conditioned medium of mouse macrophage cells was capable of mesoderm induction and, in about 30% of the cases, induced animal caps to form "embryoids," embryo-like structures with apparent axes and eyes. Eyes are highly characteristic anterior structures that had not been reported previously as the result of mesoderm induction in animal explants by *Xenopus* activin or other growth factors.

It seemed likely that the observations of Sokol and colleagues were not due to the action of a distinct inducing factor since PIF had all the properties of activin [it has since been identified as mouse activin A (Thomsen *et al.*, 1990)]. Rather, the differences more likely resided in technique. We therefore tested the two major variables in the experiment, age and size of the animal explant. Within the range of stages 7 through 9 the age of the explant did not seem to have much influence on outcome, but size of explant is very important. Animal caps subtending approximately 60° of arc [which has been standard in most laboratories (see Smith, 1987)] did not differentiate into embryoids with discernible polarity or eyes, whereas large explants produced eyes in about 20% of the cases (Figs. 3 and 4). Many of these explants also showed apparent polarity and produced cement glands, visible as dark regions at the apparent anterior end of the embryoids. In normal embryos the cement gland forms on the ventral side of the head. While cement glands are sometimes produced by small animal explants, their regular occurrence in large explants is an additional indication of anterior differentiation. Large animal explants cultured in control medium generally formed only atypical epidermis, although about 15% differentiated some mesenchyme.

The results described above appear to explain the differences between earlier observations of different authors. More importantly, they suggest that at least some animal hemisphere cells at the blastula stage are different from one another in that cells closer to the equator are capable of responding to inducer by axis formation while cells close to the pole cannot, even though they are competent to form mesodermal tissues. These cellular differences constitute a gradient (perhaps just a step function) of competence from equator to pole, establishing a "prepattern" that is not apparent until made visible by the action of an inducer (activin). Axis formation *in situ* may then depend on the cooperation between cellular differences within the blastula and the action of inducing substance(s). These differences among responding cells may themselves arise by earlier inductive influences, as suggested by earlier work (Boterenbrood and Nieuwkoop, 1973; Gimlich and Gerhart, 1984; Gimlich, 1986). The conclusions outlined above are in general agreement with results of Sokol and Melton (1991) who showed that dorsal halves of animal caps (presumably large ones) form embryoids, while ventral halves do not. The existence of differences in

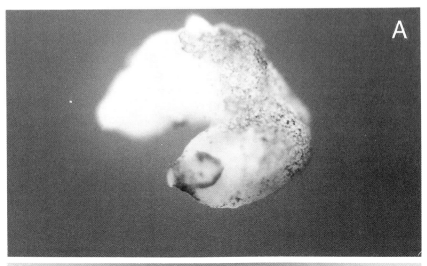

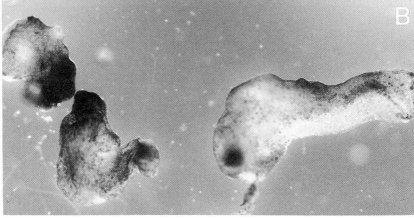

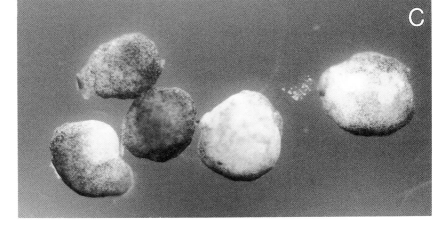

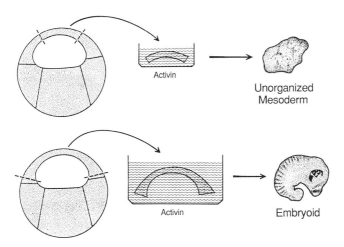

Fig. 4. Standard size (small) animal caps, subtending about 60° of arc, give rise to unorganized mesoderm with no obvious polarity when cultured for 2 to 2.5 days in activin (top). Large animal caps, as illustrated at the bottom, give rise to embryoids with eyes in about 20% of the cases.

the competence to respond to neural induction, apparently representing a prepattern, has been observed in gastrula embryos by Sharpe *et al.* (1987) and Otte *et al.* (1991).

CONSEQUENCES OF MESODERM INDUCTION: THE ACTIVATION OF REGULATORY GENES

One of the valuable approaches to the study of embryogenesis that was initiated in the past decade is the isolation of genes that are expressed in specific regions of the embryo and thus provide molecular markers for induction and differentiation events. In mesoderm induction studies, muscle-specific actin mRNA has been used extensively (Mohun *et al.*, 1984; Gurdon *et al.*, 1985; Sargent *et al.*, 1986). However, this and similar markers are characteristic of the end prod-

Fig. 3. Large animal cap explants cultured in activin A give rise in about 20% of the cases to "embryoids," structures with eyes and apparent anteroposterior and dorsoventral polarity (A,B). Uninduced large explants form near-spherical atypical epidermis (C). See Fig. 4 for illustration of large and small animal caps.

ucts of induction, representing gene expression in fully differentiated tissues. In order to isolate genes that may encode regulatory factors, a search was conducted for genes that respond very rapidly to induction, resulting in the identification of several rapidly inducible cDNAs (Rosa and Dawid, 1990). Some of these genes respond only to activin, others respond to both activin and FGF. The best-studied gene of this group, named *Mix. 1*, encodes a homeodomain protein that is expressed only during the late blastula and gastrula stages. *Mix. 1* is induced immediately following activin addition even in the absence of protein synthesis and is restricted in its distribution to the vegetal hemisphere of the embryo (Rosa, 1989). The nature of homeodomain proteins as transcription factors and the developmental importance of homoebox genes has been discussed widely (Scott *et al.*, 1989; De Robertis *et al.*, 1990, 1991; Kessel and Gruss, 1990). Because of its structure and expression pattern, *Mix. 1* is most likely involved in the differentiation of the mesoderm and, possibly, of the endoderm. In addition to *Mix. 1*, several other homeobox genes expressed in *Xenopus* embryos that are responsive to mesoderm inducers have been isolated, in particular the posterior markers *Xhox3* (Ruiz i Altaba and Melton, 1989a,b) and *XIHbox6* (Sharpe *et al.*, 1987; Cho and De Robertis, 1990). These genes are expressed somewhat later in development and it is not known whether they are direct or indirect responses to induction. A recent example of an inducible organizer-specific homeobox gene isolated in our laborabory is described below. Several interesting genes of this class have been described by Blumberg *et al.* (1991).

Regulatory genes other than homeobox genes are also activated in different regions of the mesoderm by inducing signals. The best understood of these genes is *MyoD;* the MyoD protein is a helix–loop–helix transcription factor that regulates muscle-specific genes and is capable of transforming certain cell types into muscle cells (Weintraub *et al.*, 1991). In the frog embryo, *MyoD* is expressed during gastrulation in the regions specified to form muscle (Hopwood *et al.*, 1989a). A different gene required for mesoderm formation, in particular for the formation of the notochord, is *Brachyury* (*T*), isolated originally as a mouse developmental gene (Herrmann *et al.*, 1990). The *Xenopus* homolog of the originally isolated mouse gene has been shown to be expressed in the entire mesoderm at early gastrula and to respond as an immediate early gene to mesoderm inducers (Smith *et al.*, 1991). Two additional regulatory genes isolated on the basis of homology to their *Drosophila* counterparts, *Xtwi* and *Xsna*, are expressed in the mesoderm and are inducible by activin (Hopwood *et al.*, 1989b; Sargent and

Bennet, 1990). All of these genes respond in slightly different way to induction and are expressed in distinct spatial patterns in the embryo. It may be suggested that the expression patterns of these and additional regulatory genes form the molecular basis for the differentiation of the embryo into different tissues within the body plan characteristic of the organism.

THE LIM CLASS OF HOMEOBOX GENES IN *Xenopus* DEVELOPMENT

Subclasses of homeodomain proteins have been described in which a second conserved motif is associated with the homeodomain; examples are the paired box (Bopp *et al.*, 1986) and the POU domain (Herr *et al.*, 1988). Recently, three homeobox genes have been described that share two copies of a conserved motif of cysteine and histidine residues; using the first letters of these three prototype genes the Cys/His motif has been called the LIM motif. Two of the prototype LIM genes, *mec-3* (Way and Chalfie, 1988) and *lin-11* (Freyd *et al.*, 1990) are cell lineage determinants in *Caenorhaleditis elegans*, while the *Isl-1* gene encodes a DNA-binding factor in the rat (Karlsson *et al.*, 1990). In addition, proteins of a distinct class have been described that contain one or two LIM domains but no homeodomain. Members of this class are *Ttg*/rhombotin-1 and -2 (McGuire *et al.*, 1989; Boehm *et al.*, 1990, 1991), cysteine-rich intestinal protein or CRIP (Birkenmeier and Gordon,1986), and cysteine-rich protein or CRP (Liebhaber *et al.*, 1990). Based on the wide distribution of the LIM motif and the biological importance of the two *C. elegans* LIM class homeobox genes we decided to look for homologous genes in *Xenopus*.

A polymerase chain reaction-based approach using primers within the homeobox yielded two types of sequences from *Xenopus* embryo cDNA libraries which we have named *Xlim-1* and *Xlim-3*. Both genes show substantial sequence homologies to the nematode genes *mec-3* and *lin-11*. While *Xlim-3* mRNA is expressed only in later embryogenesis and in adult stages, *Xlim-1* is expressed at early stages as well. *Xlim-1* transcripts are present at low abundance in maternal RNA and begin zygotic accumulation shortly after the midblastula transition. RNA abundance rises rapidly during gastrulation, falls in the early neurula, and subsequently rises again. We have shown (unpublished results) that the two quantitatively distinct periods of *Xlim-1* expression correspond to different phases of spatial regulation. In the gastrula embryo, *Xlim-1* RNA is localized at the dorsal lip and the advancing dorsal mesoderm, i.e., in the Spemann organizer. At later stages, expression is localized in the central nervous system and in lateral, nonsomitic mesoderm.

During normal embryogenesis, the invagination of mesodermal cells initiates at the dorsal lip and progresses along the blastocoel roof toward the animal pole, as illustrated in Fig. 5. In the process the dorsal axis of the embryo, including its neural plate, notochord, and somites, is established and organized along the anteroposterior axis. This patterning is controlled by the dorsal lip and dorsal mesoderm, as shown initially by Spemann and Mangold (1924) by transplantation into a host embryo (reviewed in Dawid *et al.*, 1990; Melton, 1991; De Robertis *et al.*, 1991). The dorsal lip not only is capable of inducing a

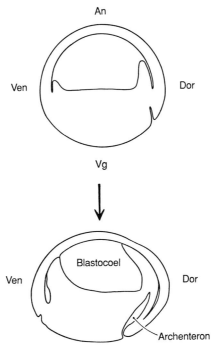

Fig. 5. Drawings of early and midgastrula embryos of *Xenopus laevis.* Animal (An), vegetal (Vg), dorsal (Dor), and ventral (Ven) are indicated. Mesoderm invagination begins at the dorsal lip (top) and proceeds along the blastocoel roof toward the animal pole (bottom). The ectoderm overlying the dorsal mesoderm forms the neural plate during this period of development, with anterior structures forming at the leading edge of the invaginating mesoderm.

series of various tissues in the host, but also elicits formation of a distinct pattern; hence its designation as the organizer. The localization of *Xlim-1* RNA in the dorsal lip and involuting dorsal mesoderm of the blastula is of great interest because the availability of specific markers for the organizer region offers the hope of analyzing its remarkable biological properties at the molecular level.

Xlim-1 is rapidly inducible by activin and, by the criterion of cycloheximide resistance, is an immediate early response to induction. *Xlim-1* RNA is inducible by retinoic acid (RA) which distinguishes it from all other genes tested in the *Xenopus* animal cap system so far; it should be noted that RA alone has no inducing effect in this system in terms of morphological changes or the differentiation of distinguishable mesodermal tissues. RA has been shown to affect frog development by deleting anterior structures: tadpoles treated with RA show a graded response, eventually leading to the absence of the entire head (Durston *et al.*, 1989). Part of the RA effect appears to be mediated by the mesoderm (Ruiz i Altaba and Jessel, 1991). Furthermore, RA increases the induction of the posterior homeobox gene *XIHbox6* by activin without being able by itself to induce this gene (Cho and De Robertis, 1990). On the basis of these results the suggestion has been made that RA mediates posterior development in the frog. Without questioning the results that form the basis of this suggestion we point to a difficulty that arises from our observations: *Xlim-1* is a gene with an early dorsal and subsequent anterior expression pattern, yet its expression is strongly stimulated by RA. Perhaps the linkage of RA with ventroposterior development is not general, and the situation must be evaluated separately for each of the presumably multiple developmental steps that can be affected by RA.

CONCLUSIONS

Embryogenesis and pattern formation in the amphibian embryo are now understood to some extent. The initial dorsoventral axis of the egg is set by a cytoplasmic rotation shortly after fertilization (Gerhart *et al.*, 1989), superimposing this pattern on the preexisting animal–vegetal axis that is established during oogenesis (see Gerhart *et al.*, 1986; Danilchik and Gerhart, 1987). The molecular basis for either of these axis specifications is not well understood, although RNA molecules have been discovered that are differentially distributed along the animal–vegetal axis of the egg (Rebagliati *et al.*, 1985; Melton, 1987). It is clear that the execution of the program that leads from the initial

specification of dorsoventral polarity to its definitive expression involves the same cell interactions that are required for mesoderm differentiation of the marginal zone (reviewed in Smith, 1989; Dawid *et al.*,
1990; Melton, 1991). The signals responsible for these cell interactions
are transmitted, at least in large measure, by soluble molecules of the
peptide growth factor class; activin and FGF are strong candidates for
inducers of mesoderm *in vivo*, and the involvement of member(s) of
the *Wnt* family is indicated (Sokol *et al.*, 1991; Smith and Harland,
1991).

Induction factors have many effects on responsive tissue, and it is
clear that the state of responsiveness, or competence, is a key element
in the nature of the response. By the midblastula stage the cells that
can respond to activin by the elaboration of mesoderm have differentiated along the animal–vegetal and dorsoventral axes, so that explants
derived from different regions of the animal hemisphere respond
differently to the same stimulus (Sokol and Melton, 1991, and experiments described above). Differential development along the major
body axes of the embryo therefore appears to be the result of interactions of inductive stimuli with differences established in the responding tissue.

Induction factors lead to both the activation and inactivation of
tissue-specific gene expression in responsive cells (Gurdon *et al.*,
1985; Dawid *et al.*, 1988; Symes *et al.*, 1988). Of particular interest is
the rapid activation of a series of genes encoding transcription factors,
since it is likely that subsequent events of differentiation and pattern
formation are controlled by early response regulatory genes. Several
homeobox genes are among the earliest transcriptional responses to
the addition of activin to competent blastula cells, making these genes
prime targets for further analysis.

REFERENCES

Asashima, M., Nakano, H., Shimada, K., Kinoshita, K., Ishii, K., Shibai, H., and Ueno, N.
    (1990). Mesodermal induction in early amphibian embryos by activin A (erythroid
    differentiation factor). *Roux's Arch. Dev. Biol.* **198**, 330–335.
Asashima, M., Nakano, H., Uchiyama, H., Sugino, H., Nakamura, T., Eto, Y., Ejima, D.,
    Nishimatsu, S.-I., Ueno, N., and Kinoshita, K. Presence of activin (erythroid differ-
    entiation factor) in unfertilized eggs and blastulae of *Xenopus laevis*. *Proc. Natl.
    Acad. Sci. U.S.A.*, **88**, 6511–6514.
Birkenmeier, E. H., and Gordon, J. I. (1986). Developmental regulation of a gene that
    encodes a cysteine-rich intestinal protein and maps near the murine immunoglobu-
    lin heavy chain locus. *Proc. Natl. Acad. Sci. U.S.A.* **83**, 2516–2520.

Blumberg, B., Wright, C. V. E., De Robertis, E. M., and Cho, K. W. Y. (1991). Organizer-specific homeobox genes in *Xenopus laevis* embryos. *Science* **253**, 194–196.

Boehm, T., Foroni, L., Kennedy, M., and Rabbitts, T. H. (1990). The rhombotin gene belongs to a class of transcriptional regulators with a potential novel protein dimerization motif. *Oncogene* **5**, 1103–1105.

Boehm, T., Foroni, L., Kaeno, M., Perutz, M. F., and Rabbitts, T. H. (1991). The rhombotin family of cysteine-rich LIM-domain oncogenes: Distinct members are involved in T-cell translocations to human chromosomes 11p15 and 11p13. *Proc. Natl. Acad. Sci.* **88**, 4367–4371.

Bopp, D., Burri, M., Baumgartner, S., Frigerio, G., and Noll, M. (1986). Conservation of a large protein domain in the segmentation gene *paired* and in functionally related genes of Drosophila. *Cell* **47**, 1033–1040.

Boterenbrood, E. C., and Nieuwkoop, P. D. (1973). The formation of mesoderm in urodelan amphibians. V. Its regional induction by the endoderm. *Wilhelm Roux Arch. Entwicklungsmech. Org.* **173**, 319–332.

Cho, K. W. Y., and De Robertis, E. M. (1990). Differential activation of *Xenopus* homeobox genes by mesoderm-inducing growth factors and retinoic acid. *Genes Dev.* **4**, 1910–1916.

Danilchik, M. V., and Gerhart, J. C. (1987). Differentiation of the animal-vegetal axis in *Xenopus laevis* oocytes. I. Polarized intracellular translocation of platelets establishes the yolk gradient. *Dev. Biol.* **122**, 101–112.

Dawid, I. B., Rebbert, M. L., Rosa, F., Jamrich, M., and Sargent, T. D. (1988). Gene expression in amphibian embryogenesis. *In* "Regulatory Mechanisms in Developmental Processes" (G. Eguchi, T. S. Okada, and L. Saxén, Eds.), p. 67–74. Elsevier, Ireland.

Dawid, I. B., Sargent, T. D., and Rosa, F. (1990). The role of growth factors in embryonic induction in amphibians. *Curr. Top. Dev. Biol.* **24**, 261–288.

De Robertis, E. M., Oliver, G., and Wright, C. V. E. (1990). Homeobox genes and the vertebrate body plan. *Sci. Am.* **263**, 46–52.

De Robertis, E. M., Morita, E. A., and Cho, K. W. Y. (1991). Gradient fields and homeobox genes. *Development* **112**, 669–678.

Durston, A. J., Timmermans, J. P. M., Hage, W. J., Hendriks, H. F. J., de Vries, N. J., Heideveld, M., and Nieuwkoop, P. D. (1989). Retinoic acid causes an anteroposterior transformation in the developing central nervous system. *Nature* **340**, 140–144.

Freyd, G., Kim, S. K., and Horvitz, H. R. (1990). Novel cystein-rich motif and homeodomain in the product of the *Caenorhabditis elegans cell lineage gene lin-11*. *Nature* **344**, 876–879.

Gerhart, J., Danilchik, M., Roberts, J., Rowning, B., and Vincent, J.-P. (1986). Primary and secondary polarity of the amphibian oocyte and egg. *In* "Gametogenesis and the Early Embryo." A. R. Liss, New York.

Gerhart, J., Danilchik, M., Doniach, T., Roberts, S., Rowning, B., and Stewart, R. (1989). Cortical rotation of the *Xenopus* egg: Consequences for the anteroposterior pattern of embryonic dorsal development. *Development* **107**(Suppl), 37–51.

Gimlich, R. L. (1986). Acquisition of developmental autonomy in the equatorial region of the *Xenopus* embryo. *Dev. Biol.* **115**, 340–352.

Gimlich, R. L., and Cooke, J. (1983). Cell lineage and the induction of second nervous systems in amphibian development. *Nature* **306**, 471–473.

Gimlich, R. L., and Gerhart, J. C. (1984). Early cellular interactions promote embryonic axis formation in *Xenopus laevis*. *Dev. Biol.* **104**, 117–130.

Gurdon, J. B., Fairman, S., Mohun, T. J., and Brennan, S. (1985). Activation of muscle-specific actin genes in Xenopus development by an induction between animal and vegetal cells of a blastula. *Cell* **41**, 913–922.

Herr, W., Sturm, R. A., Clerc, R. G., *et al.* (1988). The POU domain: A large conserved region in the mammalian *pit*-1, *oct*-1, *oct*-2, and *Caenorhabditis elegans unc*-86 gene products. *Genes Dev.* **2**, 1513–1516.

Herrmann, B. G., Labeit, S., Poustka, A., King, T. R., and Lehrach, H. (1990). Cloning of the *T* gene required in mesoderm formation in the mouse. *Nature* **343**, 617–622.

Hopwood, N. D., Pluck, A., and Gurdon, J. B. (1989a). MyoD expression in the forming somites is an early response to mesoderm induction in *Xenopus* embryos. *EMBO J.* **8**, 3409–3417.

Hopwood, N. D., Pluck, A., and Gurdon, J. B. (1989b). A *Xenopus* mRNA related to *Drosophila twist* is expressed in response to induction in the mesoderm and the neural crest. *Cell* **59**, 893–903.

Jacobson, M. (1984). Cell lineage analysis of neural induction: Origins of cells forming the induced nervous system. *Dev. Biol.* **102**, 122–129.

Karlsson, O., Thor, S., Norberg, T., Ohlsson, H., and Edlund, T. (1990). Insulin gene enhancer binding protein Isl-1 is a member of a novel class of proteins containing both a homeo- and a cys-his domain. *Nature* **344**, 879–882.

Kessel, M., and Gruss, P. (1989). Murine developmental control genes. *Science* **249**, 374–379.

Kimelman, D., and Kirschner, M. (1987). Synergistic induction of mesoderm by FGF and TGF-β and the identification of an mRNA coding for FGF in the early Xenopus embryo. *Cell* **51**, 869–877.

Köster, M., Plessow, S., Clement, J. H., Lorenz, A., Tiedemann, H., and Knöchel, W. (1991). Bone morphogenetic protein 4 (BMP-4), an member of the TGF-β family, in early embryos of *Xenopus leavis:* Analysis of mesoderm inducing activity. *Mech. Dev.* **33**, 191–200.

Liebhaber, S. A., Emery, J. G., Urbanek, M., Wang, X., and Cooke, N. E. (1990). Characterization of a human cDNA encoding a widely expressed and highly conserved custeine-rich protein with an unusual zinc-finger motif. *Nucleic Acids Res.* **18**, 3871–3879.

McGuire, E. A., Hockett, R. D., Pollock, K. M., Bartholdi, M. F., O'Brien, S. J., and Korsmeyer, S. J. (1989). The t(11;14)(p15;q11) in a T-cell acute lymphoblastic leukemia cell line activates multiple transcripts, including *Ttg-1*, a gene encoding a potential zinc finger protein. *Mol. Cell. Biol.* **9**, 2124–2132.

Melton, D. A. (1987). Translocation of a localized maternal mRNA to the vegetal pole of *Xenopus* oocytes. *Nature* **328**, 80–82.

Melton, D. A. (1991). Pattern formation during animal development. *Science* **252**, 234–241.

Mohun, T. J., Brennan, S., Dathan, N., Fairman, S., and Gurdon, J. B. (1984). Cell type-specific activation of actin genes in the early amphibian embryo. *Nature* **311**, 716–721.

Nieuwkoop, P. D. (1969). The formation of mesoderm in urodelan amphibians. I. Induction by the endoderm. *Wilhelm Roux Arch.* **162**, 341–373.

Nieuwkoop, P. D. (1973). The "organisation center" of the amphibian embryo: Its origin, spatial organization, and morphogenetic action. *Adv. Morphogenet.* **10**, 1–39.

Otte, A. P., Kramer, I. M., and Durston, A. J. (1991). Protein kinase C and regulation of the local competence of Xenopus ectoderm. *Science* **251**, 570–573.

Rebagliati, M. R., Weeks, D. L., Harvey, R. P., and Melton, D. A. (1985). Identification and cloning of localized maternal RNAs from Xenopus eggs. *Cell* **42**, 769–777.

Rosa, F. M. (1989). Mix. 1, an homeobox mRNA inducible by mesoderm inducers, is expressed mostly in the presumptive endodermal cells of Xenopus embryos. *Cell* **57**, 965–974.

Rosa, F. M., and Dawid, I. B. (1990). Molecular studies on embryonic induction. *In* "UCLA Symposium on Developmental Biology," pp. 163–174. Wiley–Liss, New York.

Rosa, F., Roberts, A. B., Danielpour, D., Dart, L. L., Sporn, M. B., and Dawid, I. B. (1988). Mesoderm induction in amphibians: the role of TGF-β2-like factors. *Science* **239**, 783–785.

Ruiz i Altaba, A., and Jessell, T. (1991). Retinoic acid modifies mesodermal patterning in early *Xenopus* embryos. *Genes Dev.* **5**, 175–187.

Ruiz i Altaba, A., and Melton, D. A. (1989a). Interaction between peptide growth factors and homeobox genes in the establishment of antero-posterior polarity in frog embryos. *Nature* **341**, 33–38.

Ruiz i Altaba, A., and Melton, D. A. (1989b). Involvement of the Xenopus homeobox gene Xhox3 in pattern formation along the anterior-posterior axis. *Cell* **57**, 317–326.

Sargent, M. G., and Bennett, M. F. (1990). Identification in *Xenopus* of a structural homologue of the *Drosophila* gene *snail*. *Development* **109**, 967–973.

Sargent, T. D., Jamrich, M., and Dawid, I. B. (1986). Cell interactions and the control of gene activity during early development of *Xenopus laevis*. *Dev. Biol.* **114**, 238–246.

Scott, M. P., Tamkun, J. W., and Hartzell, G. W., III. (1989). The structure and function of the homeodomain. *Biochim. Biophys. Acta* **989**, 25–48.

Sharpe, C. R., Fritz, A., De Robertis, E. M., and Gurdon, J. B. (1987). A homeobox-containing marker of posterior neural differentiation shows the importance of predetermination in neural induction. *Cell* **50**, 749–758.

Slack, J. M. W., Darlington, B. G., Heath, J. K., and Godsave, S. F. (1987). Mesoderm induction in early *Xenopus* embryos by heparin-binding growth factors. *Nature* **326**, 197–200.

Smith, J. C. (1987). A mesoderm-inducing factor is produced by a Xenopus cell line. *Development* **99**, 3–14.

Smith, J. C. (1989). Mesoderm induction and mesoderm-inducing factors in early amphibian development. *Development* **105**, 665–677.

Smith, W. C., and Harland, R. M. (1991). Injected Xwnt-8 RNA acts early in Xenopus embryos to promote formation of a vegetal dorsalizing center. *Cell* **67**, 753–765.

Smith, J. C., and Slack, J. M. W. (1983). Dorsalization and neural induction: Properties of the organizer in *Xenopus laevis*. *J. Embryol. Exp. Morphol.* **78**, 299–317.

Smith, J. C., Price, B. M. J., Van Nimmen, K., and Huylebroeck, D. (1990). Identification of a potent *Xenopus* mesoderm-inducing factor as a homologue of activin A. *Nature* **345**, 729–731.

Smith, J. C., Price, B. M. J., Green, J. B. A., Weigel, D., and Herrmann, B. G. (1991). Expression of a Xenopus homolog of *Brachyury (T)* is an immediate early response to mesoderm induction. *Cell* **67**, 79–87.

Sokol, S., and Melton, D. A. (1991). Pre-existing pattern in *Xenopus* animal pole cells revealed by induction with activin. *Nature* **351**, 409–411.

Sokol, S., Wong, G. G., and Melton, D. A. (1990). A mouse macrophage factor induces head structures and organizes a body axis in *Xenopus*. *Science* **249**, 561–564.

Sokol, S., Christian, J. L., Moon, R. T., and Melton, D. A. (1991). Injected Wnt RNA induces a complete body axis in Xenopus embryos. *Cell* **76**, 741–752.

Spemann, H., and Mangold, H. (1924). Über Induktion von Embryonalanlagen durch Implantation artfremder Organisatoren. *Wilhelm Roux Arch. Entwicklungsmech. Org.* **100**, 599–638.

Sudarwati, S., and Nieuwkoop, P. D. (1971). Mesoderm formation in the Anuran *Xeno-pus laevis* (Daudin). *Wilhelm Roux' Arch. Dev. Biol.* **166**, 189–204.

Symes, K., Yaqoob, M., and Smith, J. C. (1988). Mesoderm induction in *Xenopus laevis:* Responding cells must be in contact for mesoderm formation but suppression of epidermal differentiation can occur in single cells. *Development* **104**, 609–618.

Thomsen, G., Woolf, T., Whitman, M., Sokol, S., Vaughan, J., Vale, W., and Melton, D. A. (1990). Activins are expressed early in Xenopus embryogenesis and can induce axial mesoderm and anteior structures. *Cell* **63**, 485–493.

Way, J. C., and Chalfie, M. (1988). *mec-3*, a homeobox-containing gene that specifies differentiation of the touch receptor neurons in *C. elegans*. *Cell* **54**, 5–16.

Weeks, D. L., and Melton, D. A. (1987). A maternal messenger RNA localized to the vegetal hemisphere in Xenopus eggs codes for a growth factor related to TGF-β. *Cell* **51**, 861–867.

Weintraub, H., Davis, R., Tapscott, S., Thayer, M., Krause, M., Benezra, R., Blackwell, T. K., Turner, D., Rupp, R., Hollenberg, S., Zhuang, Y., and Lassar, A. (1991). The *myoD* gene family: nodal point during specification of the muscle cell lineage. *Science* **251**, 761–766.

4

Relationships between Mesoderm Induction and Formation of the Embryonic Axis in the Chick Embryo

CLAUDIO D. STERN
Department of Human Anatomy
University of Oxford
South Parks Road
Oxford OX1 3QX, United Kingdom

When a young chick embryo (blastoderm) is cut into several portions, each portion is capable of *regulation*, that is, it can give rise to a complete, albeit smaller embryo (see Spratt and Haas, 1960; Khaner *et al.*, 1985; Nieuwkoop *et al.*, 1985; Bellairs, 1986). There are, in theory, two possible interpretations of this phenomenon: The first is that cells in the embryo are pluripotent, becoming specified to their ultimate fates by cell interactions. This first interpretation explains regulation by supposing that when a fragment is separated from the embryo, cells in the fragment interact to generate the correct proportions of different cell types. The second interpretation of the phenomenon of regulation requires the embryo to consist of a fairly random mix of cells already specified to their ultimate fates, which later sort to their correct destinations. According to this, isolation of a fragment of embryo leads to sorting of the cell types to newly defined sites, but the cells do not change fates.

One problem with the second interpretation is that it does not account for our ability to construct fate maps. If embryos consist of completely random mixtures of cells of different types, there should

Cell–Cell Signaling in
Vertebrate Development

not be discrete *regions* with defined fates. However, it is possible that cells sort rather soon after becoming specified, and the mosaic arrangement of two or more cell types explains rather well the ill-defined boundaries between different regions that characterise the fate maps of most vertebrates. Nevertheless, the first interpretation is the one usually accepted; it is generally assumed that cells become determined as a result of inductive interactions with other cells, so that the correct cell types are generated at least near their ultimate locations.

GERM LAYERS AND CELL MOVEMENT PATTERNS DURING PRIMITIVE STREAK FORMATION IN THE CHICK EMBRYO

At the time of egg laying, the chick blastoderm consists of a disc, some 2 mm in diameter, comprising an inner, translucent *area pellucida* and an outer, more opaque *area opaca*. The latter region only contributes to extraembryonic structures. The first of the germ layers to be present as such is the *epiblast*, which is continuous over both *areae opaca* and *pellucida*. It is a one-cell-thick epithelium which soon becomes pseudostratified and columnar, the apices of its cells facing the albumen. From the center of this initial layer arises a second layer of cells, the *primary hypoblast* (for an explanation of the terminology see Stern, 1990). At this time, the hypoblast is no more than several unconnected islands of about 5–20 cells. However, by about 6 hr of incubation, more cells are added to it (the *secondary hypoblast*) and it becomes a loose but continuous epithelium. The source of these secondary hypoblast cells is the deep (endodermal) portion of a crescent-shaped region, the *marginal zone*, which separates *areae pellucida* and *opaca* at the future posterior (caudal) end of the embryo.

The embryo now consists of two layers: the epiblast proper, facing the albumen, from which will arise all of the embryonic tissues, and the hypoblast (primary and secondary), facing the yolk, which will give rise only to extraembryonic tissues (mainly the yolk sac stalk) although it may also contain some primordial germ cells (see Bellairs, 1986; Ginsburg and Eyal-Giladi, 1987). As the hypoblast continues to spread as a layer from posterior to anterior parts of the blastodisc, cells appear between the other two layers. These are the first cells of the *mesoderm* (Vakaet, 1984). As more of these accumulate, they coalesce to form the first axial structure of the embryo, the *primitive streak,* which makes its appearance at the posterior margin of the *area pellucida* after about 10 hr of incubation (see Bellairs, 1986 for review). Later, more mesodermal cells are recruited into the primitive streak as this structure elongates along the anteroposterior axis of the embryo (Vakaet, 1984). All the mesoderm eventually migrates out of the streak to give rise to the lateral plates, intermediate mesoderm, paraxial (somitic) mesoderm, and the axial notochord.

Soon after the primitive streak forms, some of its cells begin to insert into the hypoblast, displacing it toward the edges of the *area pellucida*. These primitive streak-derived, endodermal cells form the *definitive endoderm,* which will give rise to the lining of the gut. The elongation of the primitive streak and the expansion of the blastodisc, together with further recruitment of cells derived from the posterior marginal zone (this contribution now forming the *junctional endoblast,* which is also extraembryonic), confine the original hypoblast to a crescent-shaped region underlying the anterior portion of the *area pellucida.* This region is known as the *germinal crescent* because it contains, transiently, the primordial germ cells, which will later migrate into the gonads (see Bellairs, 1986; Ginsburg and Eyal-Giladi, 1987).

A Note on the Staging System

Stages of development after the appearance of the primitive streak are classified according to the system of Hamburger and Hamilton (1951) and use Arabic numerals. The primitive streak appears at stage 2 and reaches its full length at stage 4. Before the appearance of the primitive streak, the staging system of Eyal-Giladi and Kochav (1976) is used, in Roman numerals. The last stage in this system before the appearance of the streak is stage XIV, characterized by a full hypoblast. The hypoblast begins to form as islands at stage IX and the contribution of posterior marginal zone cells to it begins at about stage XI.

ESTABLISHMENT OF CELL DIVERSITY: EXPRESSION OF CELL-TYPE-SPECIFIC MARKERS DURING PRIMITIVE STREAK FORMATION

The apparent complexity of the processes that lead to germ layer formation and of the subsequent movements that rearrange cells leads us to address the questions: when do different embryonic cells become committed to their fates, and what are the mechanisms responsible?

To determine the time during embryonic development at which cell diversity becomes established, it is advantageous to possess cell-type-specific markers that allow different cell types to be identified as soon as possible after their appearance. Until recently, no such markers were available in the avian embryo; different cell types could only be recognized by their morphology and their position. As a result of an exhaustive search, several molecular markers for different tissues or regions of the primitive streak-stage embryo have been reported (Stern and Canning, 1988).

In our recent studies, we have concentrated mainly on the expression of immunoreactivity with monoclonal antibody HNK-1. This antibody was originally raised by Abo and Balch (1981) against a cell surface determinant on human natural killer cells. Since then, it has been used extensively as a marker for cells of the developing peripheral nervous system of avian embryos, particularly neural crest cells (e.g. Tucker *et al.*, 1984, 1988; Rickmann *et al.*, 1985; Bronner-Fraser, 1986; Stern *et al.*, 1986). At early stages of chick development, HNK-1 recognizes the cells of the hypoblast from the time they ingress from the overlying epiblast to form the islands of the primary hypoblast. It also recognizes the cells of the hypoblast and their precursors in the posterior marginal zone (Canning and Stern, 1988; Stern, 1990). The cells of the hypoblast remain HNK-1-positive even after they have become confined to the anterior germinal crescent.

As the primitive streak makes its appearance, the antibody recognizes all of the middle layer cells; these cells continue to express immunoreactivity until the end of primitive streak formation (stage 4), when some of them leave the streak region. However, just before streak formation (stages XII–XIV), a mosaic of HNK-1-positive and -negative cells is found in the epiblast; positive cells gradually become concentrated at the posterior end of the blastodisc and eventually disappear from the epiblast as the primitive streak forms. This finding led to the suggestion (Canning and Stern, 1988; Stern and Canning, 1990), which will be discussed in detail below, that the HNK-1-positive cells of the epiblast prior to streak formation are the progenitors of the (also HNK-1-positive) cells of the streak itself, which will give rise to the mesoderm and endoderm of the embryo.

At later stages of development, the notochord, splanchnopleural mesoderm (especially in cranial regions), neural crest cells, motor axons, and associated glial cells express the epitope (see Lim *et al.*, 1987; Lunn *et al.*, 1987; Canning and Stern, 1988; Tucker *et al.*,

1988). In addition, other structures, like the posterior half of the otic vesicle (Lim *et al.*, 1987; Kuratani, 1991) and portions of the developing retina and lens and of the aortic endothelium, also express immunoreactivity (unpublished observations). It has been suggested (Canning and Stern, 1988) that HNK-1 immunoreactivity characterizes cells involved in inductive interactions during development.

NATURE OF MACROMOLECULES CARRYING THE L2/HNK-1 EPITOPE AND THEIR POSSIBLE INVOLVEMENT IN GASTRULATION

The epitope recognized by the HNK-1 antibody has been studied extensively. It is a complex, sulfated carbohydrate structure (Chou *et al.*, 1986) known as L2/HNK-1 and is common to many molecules thought to play a role in cell–cell or cell–substrate adhesion (Kruse *et al.*, 1984, 1985; Keilhauer *et al.*, 1985; Bollensen and Schachner, 1987; Cole and Schachner, 1987; Hoffman and Edelman, 1987; Pesheva *et al.*, 1987; Hoffman *et al.*, 1988; Künemund *et al.*, 1988). They include neural cell adhesion molecule (NCAM), neural–glial cell adhesion molecule (NG-CAM) (also known as G4 in the chick and L1 in the mouse), a chondroitin sulfate proteoglycan receptor for cytotactin (cytotactin-binding proteoglycan; CTB-proteoglycan), the adhesion molecule on glia (AMOG), and the laminin/fibronectin receptor, integrin. It is also present on other molecules expressed in association with the nervous system such as the glycoprotein P0, myelin-associated glycoprotein (MAG), myelin basic protein (MBP), and the enzyme acetylcholinesterase (Bon *et al.*, 1987). The epitope has also been reported to form part of certain complex glycolipids (e.g., Chou *et al.*, 1986; Dennis *et al.*, 1988).

The promiscuity of the L2/HNK-1 epitope among molecules with putative roles in adhesion has led several investigators (Keilhauer *et al.*, 1985; Bronner-Fraser, 1987; Künemund *et al.*, 1988) to suggest that it may play a role in modulating the adhesive properties of its host molecules. There is, in fact, some direct evidence supporting this view: sulfated oligosaccharides contained within the epitope can disrupt cell–substrate adhesion of certain cultured cells (Künemund *et al.*, 1988). However, how this modulation is achieved is not yet understood. The epitope may also be involved in conferring specificity to cell recognition because the bonds between the component sugar groups may display some variation, as they do in blood-group-specific oligosaccharides (T. Rademacher, personal communication).

Because of this complexity, it becomes important to determine the nature of the molecules bearing the L2/HNK-1 epitope during early development of the chick embryo. Initial investigations (Canning and Stern, 1988; Canning, 1989, and unpublished) used affinity purification of detergent-solubilized proteins followed by sodium dodecyl sulfate–polyacrylamide electrophoresis (SDS–PAGE) and immunoblotting with antibodies against glycoproteins known to express the epitope.

Several bands were seen in gels run from eluted material. The most prominent is a triplet of bands at about 70 kDa, which can be obtained from samples of all germ layers. Other bands are more tissue-specific; the epiblast from stage XIV embryos shows a major unique band in excess of 300 kDa. This polypeptide is recognized by antibodies against J1/tenascin and against chondroitin sulfate (Canning, 1989, and unpublished observations). This suggests that it is related or identical to the CTB-proteoglycan, which is known to contain the L2/HNK-1 epitope (Hoffman and Edelman, 1987; Hoffman et al., 1988). Antibodies directed to J1/tenascin or to chondroitin sulfate only stain the basal lamina underlying the epiblast from stage XIV. Samples of hypoblast show a more minor specific component of about 140 kDa. This reacts with mono- and polyclonal antibodies directed to G4 (the chick equivalent of L1, or NG-CAM), which is also known to carry the L2/HNK-1 epitope (see Fushiki and Schachner, 1986). In frozen sections, antibodies to G4 stain the hypoblast and the basal lamina of the epiblast as well as cells in the posterior marginal zone. Samples of primitive streak are characterized by a band of 32 kDa, which does not change in mobility upon reduction and which is not recognized by antibodies to any of the known HNK-1-related molecules. However, antiserum raised against purified 32 kDa protein eluted from SDS–PAGE gels does not show any tissue specificity (Canning, 1989, and unpublished observations).

FATE OF HNK-1-POSITIVE CELLS

Although we do not yet know the precise functions of the L2/HNK-1 epitope, some recent experiments have used this antibody for following the fates of HNK-1-positive cells. Expression of region-specific markers is not a sufficient criterion to decide whether cells that differ in their expression of antibody immunoreactivity are developmentally different from one another. For this reason, two simple experiments were performed to establish whether the HNK-1-

positive cells found in the epiblast prior to primitive streak formation are in fact the precursors of the (HNK-1-positive) cells of the primitive streak, which in turn give rise to the mesoderm and embryonic endoderm (Stern and Canning, 1990).

Fate of HNK-1-Positive Cells of the Epiblast

In the first experiment, the HNK-1 antibody was coupled directly to colloidal gold particles. Living embryos (stage XIII) from which the hypoblast had been removed were incubated briefly in this reagent, during which time the HNK-1-positive cells become labeled; during subsequent incubation at 37°C these labeled cells endocytose the antibody–gold complex and subsequently can be recognized by the presence of gold particles in their cytoplasm. After further incubation, the fates of the cells that were HNK-1-positive at the time of labeling can be identified. It was found that gold-labeled cells are found in the primitive streak and its mesodermal and endodermal derivatives, but not in the epiblast. This result suggests that the HNK-1-positive cells of the epiblast found prior to primitive streak-derived mesendoderm.

The second experiment of Stern and Canning (1990) was designed to test whether the HNK-1-negative cells of the epiblast have the ability to contribute to the primitive streak in the absence of HNK-1-positive cells. Embryos at stage XIII (before streak formation) without their hypoblasts were incubated in a mixture of the HNK-1 antibody and guinea pig complement (Stern *et al.*, 1986) to kill the epiblast HNK-1-positive cells. Subsequent incubation of the embryo led to normal expansion of the treated embryo, but no axial structures formed. However, such embryos could be rescued by a graft of quail primitive streak cells. In this case, a normal axis formed, and the mesoderm and endoderm of such chimeric embryos was composed entirely of quail cells. This result suggests that the HNK-1-positive cells of the epiblast at stage XIII are necessary for the formation of a normal primitive streak

Fate of the HNK-1-Positive Cells of the Deep Posterior Marginal Zone

In another series of experiments (Stern, 1990), the fate of the HNK-1-positive cells of the deep posterior marginal zone was followed using the HNK-1–gold technique in combination with grafting. A donor embryo was incubated in the HNK-1–gold reagent and its deep posterior marginal zone grafted into an unlabeled recipient at the same stage of development (stage XII). The gold-labeled cells contributed to the hypoblast, but not to other tissues. When the HNK-1-positive cells of the posterior margin were ablated with a mixture of HNK-1 and complement, the grafted margin did not contribute to the hypoblast. These experiments suggest that the deep posterior marginal zone contains special, HNK-1-positive cells which are precursors of the hypoblast and that the remaining (HNK-1-negative) cells are unable to contribute to this germ layer.

MARGINAL ZONE: ORGANIZER OF THE EARLY EMBRYO

In 1933, C. H. Waddington (see also Azar and Eyal-Giladi, 1981) rotated the hypoblast of a prestreak chick embryo to reverse its anteroposterior orientation. He found that the primitive streak now formed from the opposite (anterior) margin of the blastodisc and proposed that the hypoblast induces the primitive streak.

Spratt and Haas (1960) demonstrated the importance of the posterior marginal zone in the formation of the embryonic axis, based on a study of the regulative ability of isolated pieces of blastoderm. They postulated the existence of a gradient in embryo-forming potential along the circumference of the marginal zone ring, with its highest point at the posterior margin. Since then, Eyal-Giladi and colleagues have expanded on this suggestion (Eyal-Giladi and Spratt, 1965; Azar and Eyal-Giladi, 1979; Mitrani *et al.*, 1983; Khaner *et al.*, 1985; Khaner and Eyal-Giladi, 1986, 1989; Eyal-Giladi and Khaner, 1989); they found that rotation of the marginal zone about its anteroposterior axis resulted in reversal of the orientation of the streak, as Waddington had found after hypoblast rotation. They also suggested that the posterior marginal zone acts both to induce the primitive streak and to prevent the formation of secondary axes (Eyal-Giladi and Khaner, 1989).

However, the epiblast of the early chick embryo can give rise to mesoderm cells even in the absence of both hypoblast and marginal zone (Mitrani and Shimoni, 1990; Stern, 1990). It therefore seems unlikely that the interaction between marginal zone or hypoblast and epiblast represents an *induction* of *mesoderm* by these tissues as defined by Gurdon (1987). It is clear, nevertheless, that the marginal zone is required both for the formation of a primitive streak with normal morphology which undergoes normal elongation and to prevent the formation of secondary axes (Eyal-Giladi and Khaner, 1989; Khaner and Eyal-Giladi, 1989), but not for the differentiation of mesodermal cells. These observations emphasize the importance of distinguishing between *induction of the mesoderm* and *induction of the primitive streak*, a distinction that has not always been made in the literature of either amniotes or amphibians.

INDUCTION OF THE PRIMITIVE STREAK: A REVISED VIEW

To accommodate the above findings, I suggest a new scheme for the formation of the primitive streak in the chick embryo. The following experimental conclusions have to be accounted for in this model: (a) the epiblast at stage XII–XIII will give rise to mesoderm cells in the absence of hypoblast or marginal zone; (b) rotation of the marginal zone or of the hypoblast results in reversal of the anteroposterior orientation of the primitive streak and resulting embryo; (c) HNK-1-positive cells are the precursors of cells of the primitive streak, at least at early stages of development, and are required for the formation of this structure; (d) the deep posterior marginal zone contributes to the hypoblast and junctional endoblast from the HNK-1-positive cells of its deep layer and to the epiblast portion of the streak from the epiblast layer; (e) ingression of presumptive mesodermal cells into the primitive streak occurs in two stages: first as a distributed polyingression in a large region of the blastoderm and then as part of localized invagination of epiblast at the primitive streak (see Vakaet, 1984).

Based on these premises, I suggest that primitive streak formation occurs in three stages, and that this axial structure contains cells derived from three different sources. First, HNK-1-positive cells, which appear as a randomly distributed population in the epiblast, ingress individually or in small groups into the interior of the blastoderm, starting at stages XII–XIII. In the second stage (stages XIV–3), these cells accumulate at the posterior margin of the blastoderm, where they interact with the overlying epiblast. I propose that this interaction is required in order to allow the cells of the epiblast of this region to undergo a process analogous to the *convergent extension* of the amphibian marginal zone cells (Keller and Danilchik, 1988) which leads to elongation of the primitive streak. When the cells of the epiblast portion of the streak, derived from the posterior margin, have reached maximum elongation along the future anteroposterior axis (stage 3^+), the third and final stage begins: a groove develops in the center of this structure and this is accompanied by the dissolution of the basement membrane underlying this region of epiblast. This allows further cells (many of which must be HNK-1-negative, because at this stage almost no positive cells remain in the epiblast) to be recruited into the primitive streak, where they will mix with the earlier HNK-1-positive cells and will contribute, together, to the embryonic mesoderm and definitive endoderm.

However, the above model does not explain three features: first, it fails to provide a mechanism for the convergence of the early HNK-1-positive epiblast cells toward the posterior marginal zone. Second, it does not account for the effects of hypoblast rotation on the orientation of the primitive streak. Finally, it does not explain why or how some cells are allocated as HNK-1-positive, early streak precursors while others are not. The following discussion addresses these three issues.

HNK-1-Positive Cells Find the Posterior Marginal Zone: Evidence for Chemotaxis

Given that the epiblast contains a seemingly random mixture of HNK-1-positive and -negative cells, in which the former are the precursors of the primitive streak-derived mesoderm and endoderm, how do these primitive streak cells find their destination in the embryo? One possibility is that one of the roles of the posterior marginal zone is to direct the migration of the HNK-1-positive cells to its proximity, where the primitive streak starts to form. To test this hypothesis, a collagen gel assay was used to find out if chemotaxis is involved (Stern and Jephcott, unpublished results).

A small explant of central epiblast cells was placed in the center of a collagen gel matrix and confronted on one side with an explant of posterior marginal tissue and on the other with a piece of anterior marginal tissue. It was found that cells of the central epiblast migrated toward the posterior marginal explant, suggesting that the posterior marginal zone emits a chemoattractant.

It is important to determine that it is the HNK-1-positive cells and not others that respond to this chemotactic signal. To address this, we prelabeled the central epiblast explant using HNK-1–gold complex as described above and examined the cultures using a confocal scanning laser microscope. It was found that HNK-1-positive cells displayed chemoattraction to the posterior margin, while HNK-1-negative cells spread evenly away from the explant.

In an attempt to establish the molecular nature of the chemotactic signal, we incorporated heparin immobilized on agarose beads into the collagen gel. This did not prevent the migration of cells from the epiblast explant, but migration was no longer directed toward the posterior margin explant. This result suggests that the chemoattractant(s) are heparin-binding molecules; it also provides an explanation for the results of Mitrani and colleagues (Mitrani and Shimoni, 1990; Mitrani et al., 1990), in which it was suggested that heparin-binding molecules (e.g., basic-FGF, activin A) play a role in the inductive effects of the posterior marginal zone.

Role of the Hypoblast in Anteroposterior Polarity of the Embryo

The hypoblast may play a role in the process discussed above and contribute to the guidance of HNK-1-positive premesendodermal cells to the posterior marginal zone. It is possible, for example, that the chemoattractant binds to its extracellular matrix, as several heparin-binding factors are known to do. Rotation of the hypoblast, which will include the gradient of attractant, will therefore, result in migration of HNK-1-positive cells in reversed direction. Thus, rotation of the hypoblast (gradient) and rotation of the marginal zone (source of these chemicals) will have the same effects.

The Origins of Cell Diversity

The experiments that have been conducted to date on the early chick embryo cannot address the question of how the HNK-1-positive cells of the epiblast become different from their HNK-1-negative counterparts, because once they can be recognized they have already diversified. However, we might speculate on the type of mechanism that could lead to a constant proportion of cells, within a large population, to become different from the rest irrespective of their position in the embryo. In principle, there are two main ways in which this could be achieved: by local cell interactions, or by some intrinsic, preexisting diversity between the cells.

Local cell interactions could lead to diversification, if, for example, random cells within the population became different at some particular time in development, and then these cells inhibited their neighbors from undergoing a similar change. By limiting the range of such an inhibition, the proportions of the two cell types can be controlled. This mechanism will also prevent cells of the divergent cell type from forming large clusters and would distribute them evenly in the embryo.

For the second mechanism we have to assume some parameter that differs between different cells of the population at the time of diversification. An obvious candidate is the cell cycle. Suppose, for example, that at some time in development, a signal is produced (perhaps by the marginal zone, or by the hypoblast, or even by all cells) which is an inductor of mesoderm and which diffuses easily all over the embryo. Activin might be such an inducer (Mitrani *et al.*, 1990; Cooke and Wong, 1991). However, competence to respond at any one time is restricted to a subpopulation of cells, for example, to those that are in one particular phase of the cell cycle.

Within a large population of cells (the early blastoderm contains about 20,000 cells), both of the mechanisms proposed above will regulate perfectly for the size of the embryo, even within very large limits. I suggest that most vertebrate embryos will be found to contain restrictions to the competence of cells for inductive signals and that it is these restrictions that are primarily responsible for embryonic size regulation during mesoderm induction.

ACKNOWLEDGMENT

My research on this subject is funded by the Wellcome Trust.

REFERENCES

Abo, T., and Balch, C. M. (1981). A differentiation antigen of human NK and K cells identified by a monoclonal antibody (HNK-1). *J. Immunol.* **127**, 1024–1029.

Azar, Y., and Eyal-Giladi, H. (1979). Marginal zone cells: The primitive streak-inducing component of the primary hypoblast in the chick. *J. Embryol. Exp. Morphol.* **52**, 79–88.

Azar, Y., and Eyal-Giladi, H. (1981). Interaction of epiblast and hypoblast in the formation of the primitive streak and the embryonic axis in the chick, as revealed by hypoblast rotation experiments. *J. Embryol. Exp. Morphol.* **61**, 133–144.

Bellairs, R. (1986). The primitive streak. *Anat. Embryol.* **174**, 1–14.

Bollensen, E., and Schachner, M. (1987). The peripheral myelin glycoprotein P0 expresses the L2/HNK-1 and L3 carbohydrate structures shared by neural cell adhesion molecules. *Neurosci. Lett.* **82**, 77–82.

Bon, S., Meflah, K., Musset, F., Grassi, J., and Massoulie, J. (1987). An IgM monoclonal antibody, recognizing a subset of acetylcholinesterase molecules from electric organs of Electrophorus and Torpedo, belongs to the HNK-1 anti-carbohydrate family. *J. Neurochem.* **49**, 1720–1731.

Bronner-Fraser, M. (1986). Analysis of the early stages of trunk neural crest cell migration in avian embryos using monoclonal antibody HNK-1. *Dev. Biol.* **115**, 44–55.

Bronner-Fraser, M. (1987). Perturbation of cranial neural crest migration by the HNK-1 antibody. *Dev. Biol.* **123**, 321–331.

Canning, D. R. (1987). Ph.D. thesis, University of Oxford.

Canning, D. R., and Stern, C. D. (1988). Changes in the expression of the carbohydrate epitope HNK-1 associated with mesoderm induction in the chick embryo. *Development* **104**, 643–656.

Chou, D. K., Ilyas, A. A., Evans, J. R., Costello, C., Quarles, R. H., and Jungalwala, F. B. (1986). Structure of sulfated glucuronyl glycolipids in the nervous system reacting with HNK-1 antibody and some IgM paraproteins in neuropathy. *J. Biol. Chem.* **261**, 11717–11725.

Cole, G. J., and Schachner, M. (1987). Localization of the L2 monoclonal antibody binding site on chicken neural cell adhesion molecule (NCAM) and evidence for its role in NCAM-mediated adhesion. *Neurosci. Lett.* **78**, 227–232.

Cooke, J., and Wong, A. (1991). Growth-factor-related proteins that are inducers in early amphibian development may mediate similar steps in amniote (bird) embryogenesis. *Development* **111**, 197–212.

Dennis, R. D., Antonicek, H., Weigandt, H., and Schachner, M. (1988). Detection of the L2/HNK-1 carbohydrate epitope on glycoproteins and acidic glycolipids of the insect Calliphora vicina. *J. Neurochem.* **51**, 1490–1496.

Eyal-Giladi, H., and Khaner, O. (1989). The chick's marginal zone and primitive streak formation. II. Quantification of the marginal zone's potencies: Temporal and spatial aspects. *Dev. Biol.* **49**, 321–337.

Eyal-Giladi, H., and Kochav, S. (1976). From cleavage to primitive streak formation: a complementary normal table and a new look at the first stages of the development of the chick. *Dev. Biol.* **49**, 321–337.

Eyal-Giladi, H., and Spratt, N. T. (1965). The embryo forming potencies of the young chick blastoderm. *J. Embryol. Exp. Morphol.* **13**, 267–273.

Fushiki, S., and Schachner, M. (1986). Immunocytological localization of cell adhesion molecules L1 and N-CAM and the shared carbohydrate epitope L2 during development of the mouse neocortex. *Dev. Brain. Res.* **24**, 153–167.

Ginsburg, M., and Eyal-Giladi, H. (1987). Primordial germ cells of the young chick blastoderm originate from the central zone of the area pellucida irrespective of the embryo-forming process. *Development* **101**, 209–220.

Gurdon, J. B. (1987). Embryonic induction—molecular prospects. *Development* **99**, 285–306.

Hamburger, V., and Hamilton, H. L. (1951). A series of normal stages in the development of the chick. *J. Morphol.* **88**, 49–92.

Hoffman, S., and Edelman, G. (1987). A proteoglycan with HNK-1 antigenic determinants is a neuron associated ligand for cytotactin. *Proc. Natl. Acad. Sci. U.S.A.* **84**, 2523–2527.

Hoffman, S., Crossin, K. L., and Edelman, G. M. (1988). Molecular forms, binding functions and developmental expression patterns of cytotactin and cytotactin binding proteoglycan, an interactive pair of extracellular matrix molecules. *J. Cell Biol.* **106**, 519–532.

Keilhauer, G., Faissner, A., and Schachner, M. (1985). Differential inhibition of neurone-neurone, neurone-astrocyte and astrocyte-astrocyte adhesion by L1, L2, and N-CAM antibodies. *Nature* **316**, 728–730.

Keller, R., and Danilchik, M. (1988). Regional expression, pattern and timing of convergence and extension during gastrulation in Xenopus laevis. *Development* **103**, 193–209.

Khaner, O., and Eyal-Giladi, H. (1986). The embryo-forming potency of the posterior marginal zone in stages X through XII of the chick. *Dev. Biol.* **115**, 275–281.

Khaner, O., and Eyal-Giladi, H. (1989). The chick's marginal zone and primitive streak formation. I. Coordinative effect of induction and inhibition. *Dev. Biol.* **134**, 206–214.

Khaner, O., Mitrani, E., and Eyal-Giladi, H. (1985). Developmental potencies of area opaca and marginal zone areas of early chick blastoderms. *J. Embryol. Exp. Morphol.* **89**, 235–241.

Kruse, J., Mailhammer, R., Wernecke, H., Faissner, A., Sommer, I., Goridis, C., and Schachner, M. (1984). Neural cell adhesion molecules and myelin-associated glycoprotein share a carbohydrate moiety recognized by monoclonal antibodies L2 and HNK-1. *Nature* **311**, 153–155.

Kruse, J., Keilhauer, G., Faissner, A., Timpl, R., and Schachner, M. (1985). The J1 glycoprotein—a novel nervous system cell adhesion molecule of the L2/HNK-1 family. *Nature* **316**, 146–148.

Künemund, V., Jungalwala, F. B., Fischer, G., Chou, D. K. H., Keilhauer, G., and Schachner, M. (1988). The L2/HNK-1 carbohydrate of neural cell adhesion molecules is involved in cell interactions. *J. Cell. Biol.* **106**, 213–223.

Kuratani, S. C. (1991). Alternate expression of the HNK-1 epitope in rhombomeres of the chick embryo. *Dev. Biol.* **144**, 215–219.

Lim, T. M., Lunn, E. R., Keynes, R. J., and Stern, C. D. (1987). The differing effects of occipital and trunk somites on neural development in the chick embryo. *Development* **100**, 525–534.

Lunn, E. R., Scourfield, J., Keynes, R. J., and Stern, C. D. (1987). The neural tube origin of ventral root sheath cells in the chick embryo. *Development* **101**, 247–254.

Mitrani, E., and Shimoni, Y. (1990). Induction by soluble factors of organized axial structures in chick epiblasts. *Science* **247**, 1092–1094.

Mitrani, E., Shimoni, Y., and Eyal-Giladi, H. (1983). Nature of the hypoblastic influence on the chick embryo epiblast. *J. Embryol. Exp. Morphol.* **75**, 21–30.

Mitrani, E., Gruenbaum, Y., Shohat, H., and Ziv, T. (1990). Fibroblast growth factor during mesoderm induction in the early chick embryo. *Development* **109**, 387–393.

Nieuwkoop, P. D., Johnen, A. G., and Albers, B. (1985). "The Epigenetic Nature of Early Chordate Development." Cambridge Univ. Press.

Pesheva, P., Horwitz, A. F., and Schachner, M. (1987). Integrin, the cell surface receptor for fibronectin and laminin, expresses the L2/HNK-1 and L3 carbohydrate structures shared by adhesion molecules. *Neurosci. Lett.* **83**, 303–306.

Rickmann, M., Fawcett, J. W., and Keynes, R. J. (1985). The migration of neural crest cells and the outgrowth of motor axons through the rostral half of the chick somite. *J. Embryol. Exp. Morphol.* **90**, 437–455.

Spratt, N. T., and Haas, H. (1960). Integrative mechanisms in development of early chick blastoderm. I. Regulative potentiality of separated parts. *J. Exp. Zool.* **145**, 97–137.

Stern, C. D. (1990). The marginal zone and its contribution to the hypoblast and primitive streak of the chick embryo. *Development* **109**, 667–682.

Stern, C. D., and Canning, D. R. (1988). Gastrulation in birds: A model system for the study of animal morphogenesis. *Experientia* **44**, 61–67.

Stern, C. D., and Canning, D. R. (1990). Origin of cells giving rise to mesoderm and endoderm in chick embryo. *Nature* **343**, 273–275.

Stern, C. D., Sisodiya, S. M., and Keynes, R. J. (1986). Interactions between neurites and somite cells: Inhibition and stimulation of axon outgrowth in the chick embryo. *J. Embryol. Exp. Morphol.* **91**, 209–226.

Tucker, G. C., Aoyama, H., Lipinski, M., Tursz, T., and Thiery, J. P. (1984). Identical reactivity of the monoclonal antibodies HNK-1 and NC-1: Conservation in vertebrates on cells derived from the neural primordium and on some leucocytes. *Cell Diff.* **14**, 223–230.

Tucker, G. C., Delarue, M., Zada, S., Boucaut, J. C., and Thiery, J. P. (1988). Expression of the HNK-1/NC-1 epitope in early vertebrate neurogenesis. *Cell Tissue Res.* **251**, 457–465.

Vakaet, L. (1984). The initiation of gastrular ingression in the chick blastoderm. *Am. Zool.* **24**, 555–562.

Waddington, C. H. (1933). Induction by the endoderm in birds. *Wilhelm Roux Arch. Entwicklungsmech. Org.* **128**, 502–521.

PART III

CELL MIGRATION
AND DIFFERENTIATION

5

Studies of Neural Cell Lineages Using Injectable Fluorescent Tracers

RICHARD WETTS AND SCOTT E. FRASER
Division of Biology, California Institute of Technology, Pasadena, California 91125

INTRODUCTION

A central question of developmental biology focuses on how the different cell types are formed during embryogenesis. As each of the specialized cell types of the adult organism has a specific function, each must be formed at the proper time and in the proper numbers to produce a healthy, functioning animal. The proper formation of these many cell types is thought to involve specific developmental events, including the proliferation of precursor cells, and specification of cell fate, and terminal differentiation. The cellular and molecular mechanisms that regulate these key developmental events are poorly understood. One approach to elucidating these mechanisms begins with the study of cell lineages, the sequence of cell divisions that produces the postmitotic, differentiated cell types. Cell lineages indicate which cell types share a common ancestor and when during development this ancestor is present.

Cell lineage analysis is an essential first step in the study of the formation of specific cell numbers and cell types. However, an understanding of the mechanisms that control the cell lineages requires further experimental analysis. For example, an unipotent clone (a clone in which all members are of the same phenotype) might reflect a restriction of the precursor's potential; however, such a clone might have resulted from an uncommitted precursor whose progeny all happened to differentiate into the same cell type. This outcome can occur as a result of any one of several different events, such as physical

67

barriers that restrict the dispersal of the descendants to a small area, thereby exposing each of them to the same extrinsic determinants. To illustrate the complexity of this issue further, a multipotent clone (a clone in which the members formed multiple phenotypes) does not necessarily indicate the absence of strict lineage control; rather, the precursor could have been committed to forming a specific set of cells with more than one phenotype. In each of these cases, the state of commitment of the precursor cell can be determined only by experimental manipulations, such as transplanting the precursor cell to a new location. The new location challenges the lineage to produce new cell types, thereby permitting the experimenter to distinguish between an uncommitted precursor (which will produce a new lineage) and a committed one (which will not). Perturbation experiments of this type can be designed and interpreted only with a knowledge of the normal cell lineages.

FLUORESCENT DEXTRAN AND RETROVIRAL VECTORS AS LINEAGE TRACERS

The study of cell lineages requires a means to identify those cells which share a common ancestor. Unfortunately, most vertebrate embryos are relatively inaccessible during development and are optically opaque; thus, individual cells cannot be observed directly. Furthermore, vertebrate embryos possess a large number of precursor cells that are not uniquely identifiable. Thus, the same precursor cannot be recognized and followed in different individuals. To overcome these difficulties such that lineages can be analyzed in vertebrate embryos, one of the precursor cells must be rendered unique so that, later in development, its descendants can be identified.

Fluorescent dextrans (Gimlich and Braun, 1985) are excellent cell lineage tracers. These vital dyes are introduced into single precursor cells by microinjection. After microinjection, the hydrophilic dextran diffuses throughout the cytoplasm of the injected cell, but the large molecular weight of the dextran molecules confines them to the injected cell. Because the dye passes to the daughter cells during cytokinesis, the only labeled cells seen later in development are the descendants of the injected cells. The fluorescent moiety allows the labeled cells to be observed in live animals (Kimmel and Warga, 1986; Warga and Kimmel, 1990) as well as in fixed tissue. In addition, different fluorescent moieties can be used to label neighboring lineages uniquely (Sheard and Jacobson, 1987; Wetts and Fraser, 1989). A major advantage of fluorescent dextran is that it can be injected into and visualized within practically any cell type of any species, thereby avoiding the species restrictions of some genetic markers. Although the injection procedure itself can be technically difficult, it allows the experimenter to control exactly the time of labeling and to choose the position of the cell to be labeled. Furthermore, visualization of the injected cell immediately after microinjection assures that only a single precursor (or the daughters of a precursor that recently had undergone cytokinesis) has been labeled. Unfortunately, only a finite amount of the dye can be injected; therefore, if a precursor is injected too early in the lineage, the dextran can be diluted by mitotic activity until it is no longer visible. Our experience has shown that this potential difficulty can be minimized by proper experimental design and by conservative interpretation of the results (Bronner-Fraser and Fraser, 1988, 1989; Fraser *et al.*, 1990; Stern *et al.*, 1988, 1991; Wetts and Fraser, 1988; Wetts *et al.*, 1989).

Another powerful approach relies on the use of retroviral vectors as lineage tracers (Sanes, 1989). A precursor is labeled by infection with an engineered virion that contains a marker gene, typically the *Escherichia coli* β-galactosidase gene. Because this marker gene is substituted for the viral replication genes, the virus is competent to integrate into the genome of the infected cell but incompetent to form new virions. Thus, the marker gene carried by the virus cannot spread to cells other than the descendants of an originally infected cell. The members of a cell lineage can later be recognized by the presence of the marker gene product, which is detected with either histochemistry or immunocytochemistry. The major advantages of this approach are the lack of dilution of the marker and the ease of labeling. By exposing the target tissue to a solution containing the virions, the natural cell–virus interaction brings the viral genome into the cell and integrates the marker gene into the genome of the infected cell. After integration, the marker gene replicates with the host genome, eliminating any possibility of dilution of the marker. Unfortunately, the use of retroviruses as lineage tracers has several disadvantages. First, the marker gene might not be expressed in all cell types; in this situation, the full potential of the precursor cannot be ascertained, because some of its descendants are unlabeled and hence are not identified as part of the clone. Second, viral infection and marker gene expression can be species-specific, necessitating a reengineering of the vector for use in different species. Finally, the number of infections cannot be controlled directly, so that a group of labeled cells that appears to be a clone might actually be descended from more than one precursor. This problem is minimized by using low titers of virus, allowing statistical analyses to be used to argue that only one precursor was infected. In addition, by studying regions of the nervous system in which clonal descendants remain together and do not intermix with other clones, the physical proximity of the labeled cells can be used to delineate the clones. In practice, these potential problems have not been restrictive, and a great deal of important data on cell lineages has been obtained with retroviral vectors (e.g., Austin and Cepko, 1990; Galileo *et al.*, 1990; Gray and Sanes, 1991; Leber *et al.*, 1990; Luskin *et al.*, 1988; Price and Thurlow, 1988; Turner *et al.*, 1990; Walsh and Cepko, 1988).

Regardless of the lineage tracer used, the strategy of cell lineage studies is relatively straightforward, starting with cell labeling and followed by, after time for development, analysis of the phenotypes and positions of the descendants. The first step in the analysis is to examine the precursor cell immediately after labeling. This provides information about the morphology and location of the labeled precursor cells, and it verifies that the labeling procedure is marking only a single cell. Note that this examination of labeled precursors may be practical only with injectable lineage tracers; retroviral vectors require time for integration and expression, preventing an immediate examination. The second step is to examine the clones of labeled cells at some time after labeling to determine their positions, numbers, and phenotypes. The range of cell types that is observed provides important information about the normal developmental potential of the precursor cells. Of course, the variety of phenotypes produced by any single precursor is a minimum estimate of its potential. The actual potential might be greater for a variety of reasons: (i) the precursor may have divided only a few times, such that the number of descendants is smaller than the possible number of cell types and making it impossible for all cell types to be represented; (ii) the descendants dispersed to occupy only a subset of the potential sites, resulting in the absence of those cell types formed in the missed sites (iii) some of the descendants were eliminated by cell death before the stage of analysis; (iv) the lineage tracer was lost in some of the descendants by lack of expression or dilution. Similarly, counts of the number of labeled descendants can provide only a lower estimate of the number of divisions that occurred. Because cell death or loss of the lineage marker would decrease the observed clone size, cell counts can only estimate the number of mitoses and indicate roughly the precursor's position in the cell lineage. In light of these caveats, it would be unwise and incorrect to assume that *all* of the descendants of the labeled precursor were identified. The cells that are distinctly labeled, however, provide sufficient information on which to base valid conclusions.

CELL LINEAGES IN THE FROG EYEBUD

The retina of the frog *Xenopus laevis* provides several advantages for cell lineage studies; its structure is relatively simple, its development is rapid, and its location is accessible for microinjection. The anatomy of the frog retina is similar to the anatomy of other vertebrate retinas (Adler and Farber, 1986; Dowling, 1987). As shown in Fig. 1, the outer nuclear layer (ONL) consists of the photoreceptors; the inner nuclear layer (INL) consists of the interneurons (the horizontal, bipolar, and amacrine cells) and the Muller glial cells; and the ganglion cell layer (GCL) consists of the ganglion cells. The small number of cell types and their laminar distribution are advantageous for recognizing the cell types of clonally related cells. Developmentally, the retina begins as an evagination of the diencephalon called the optic vesicle (St 20–St 25 in *Xenopus;* staging according to Nieuwkoop and Faber (1956)). The optic vesicle reshapes itself into the two-layered optic cup by the invagination of the most distal part of the vesicle (St 26–St 35). The

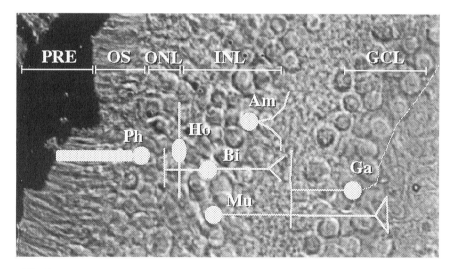

Fig. 1. Laminar organization of the neural retina. Schematic drawings of the six major cell types are shown superimposed onto a photomicrograph of a section of frog retina. The photoreceptor somas are located in the ONL; the somas of the bipolar, horizontal, amacrine, and Muller cells are located in the INL; and the ganglion cell bodies are located in the GCL. The characteristic location of each cell type aids in recognizing into which cell type that a labeled descendant has differentiated. Am, amacrine neuron; Bi, bipolar neuron; Ga, ganglion cell; GCL, ganglion cell layer; Ho, horizontal cell; INL, inner nuclear layer; Mu, Muller glial cell; ONL, outer nuclear layers; OS, outer segments of the photoreceptors; Ph, photoreceptor; PRE, pigmented retinal epithelium.

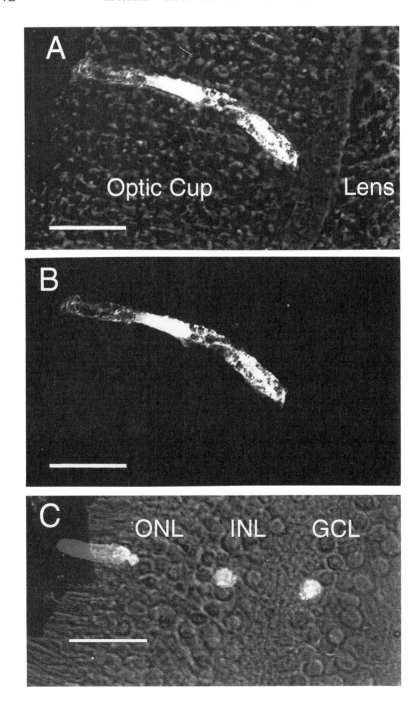

layer of the optic cup opposed to the lens primordium will differentiate into the neural retina; the layer furthest from the lens will form the pigmented retinal epithelium. Cells in the central part of the nascent neural retina begin to withdraw from the mitotic cycle as this morphogenesis is underway, between St 25 and St 37 (Beach and Jacobson, 1979; Holt *et al.*, 1988; Jacobson, 1968). The cells of the optic cup begin to differentiate into recognizable neurons shortly thereafter (St 39–St 45), resulting in a functional neural retina within 4 days of fertilization.

To determine the lineages of the neuroepithelial cells of the eye primordium, we microinjected rhodamine dextran lineage tracer into optic cup cells at St 30 (Figs. 2A and 2B). After 2–7 days (at approximately St 46–St 49), clones of labeled cells were visible in histological sections of the retina. Most of the clones (27 of 56 clones) had descendants in all three layers of the retina (Fig. 2C). Given the characteristic distribution of cell types in the retina, this multilaminar dispersal strongly suggested that these precursor cells were multipotent. The descendants were distributed in a radial column, with little spread circumferentially within the plane of the retina. This lack of cell mixing in the retina suggests that the cells differentiate very near to the site they were born.

Fig. 2.   Cells in the frog retina labeled with rhodamine dextran. (A) Rhodamine dextran was microinjected into a single precursor cell in the optic cup (St 30); immediately after microinjection, the animal was fixed, embedded, and sectioned. Fluorescence and bright-field pictures were obtained with a laser-scanning confocal microscope (Bio-Rad Richmond, CA, MRC 600 mounted on a Zeiss Axiovert) and were combined using image processing software (Adobe PhotoShop, Macintosh Version). This cell illustrates the characteristics seen in essentially all retinal precursor cells: it is undifferentiated, large, and columnar, extending across the entire width of the optic cup. (B) The same precursor cell as in (A), visualized with fluorescence only. The hydrophilic fluorescent dextran had diffused throughout the cytoplasm and nucleus but did not enter the hydrophobic lipid inclusions. These lipid stores provide nutrients during the embryonic stages of amphibian development. (C) Several days after microinjection, the rhodamine dextran identifies a clone of cells descended from a single labeled precursor. This clone is composed of three cells, one in each of the three layers of the neural retina. Since different cell types are located in each of these layers, the multilaminar distribution of the descendants indicates that their precursor produced multiple cell types. The small number of cells indicates that different cell types were produced as late as the last divisions of this lineage. The three labeled cells form a radial column across the width of the neural retina, indicating that there is little cell mixing during retinal development. The multilaminar distribution and the radial alignment seen in this clone are characteristics of many retinal clones. Abbreviations as in Fig. 1. Bars: approximately 40 μm.

In some cases, the rhodamine dextran filled the cells' processes, and specific cell types could be recognized by their characteristic morphology (Dowling, 1976; Ramon y Cajal, 1972). Precursor cells in the frog optic cup (St 30) gave rise to clones of labeled cells that contained multiple cell types (unpublished observations). Many clones included both neurons and Muller glial cells. Even the smallest clones showed great diversity; frequently, each cell of such clones was of a different cell type (e.g., Fig. 2C). The multipotency observed in these small clones suggests that frog retinal precursor cells remain uncommitted to their cell phenotype until as late as their last mitosis. The multipotency of retinal precursors has been observed independently using horseradish peroxidase in the frog (Holt *et al.*, 1988) and using retroviral vectors in rodents (Turner and Cepko, 1987; Turner *et al.*, 1990) and birds (Fekete *et al.*, 1990). In each of these studies, some clones as small as two cells contained different phenotypes; thus, it is likely that the cellular and molecular mechanisms that specify cell fate occur late in retinal development, perhaps even after the last mitosis.

To gain information about the regulation of cell division, the numbers of descendants in each clone were determined. Optic cup cells, injected at St 30, produced clones of 1–35 cells (mean, 10.8 cells; SD, 8.9; N, 56 clones). Although all 56 clones resulted from precursors labeled at essentially the same stage, the number of descendants in any one clone varied wildly within this relatively broad range. This variety in clone sizes is qualitatively similar to that of clones derived from St 22 precursor cells in the frog (Wetts and Fraser, 1988) and from precursors in the rodent optic cup labeled with a recombinant retrovirus (Turner and Cepko, 1987; Turner *et al.*, 1990). This variety in clones sizes suggests that cell proliferation is not *tightly* regulated.

To build on our lineage experiments demonstrating the wide variety of clone sizes, we have begun to explore the mechanisms that control clone size. The frog eyebud is an advantageous system for these studies, because it can heal and develop normally after many types of experimental manipulation. These experimental manipulations alter the local environment of the precursor, thereby challenging the cell lineage and revealing the role of various mechanisms. For example, to test the role of total cell number in the regulation of proliferation, we have determined clone sizes after ablating a portion of the eyebud. At approximately St 30, more than half of one eyebud was removed with sharpened forceps, and a single cell located in the remaining half was filled with rhodamine dextran lineage tracer. In control animals, an incision was made in the eyebud to control for wound healing effects, but no tissue was removed. Following surgery, both the experimental and the control eyes healed and formed structurally normal retinas.

After 2–3 days (approximately St 44), morphometric measurements confirmed that the experimental eye was reduced in size (43% of the contralateral eye), while the sham-operated eyes were normal in size (97%). Despite this large asymmetry, the mean number of rhodamine dextran-labeled descendants was essentially identical in experimental (mean, 7.7; SEM, 1.1; N, 26) and control (mean, 8.0; SEM, 1.0; N, 25) retinas. The absence of a statistically significant difference suggests that the eyebud cells did not compensate for the loss of neighboring cells and that cell number is not regulated by the total size of the tissue.

CELL LINEAGES IN THE SPINAL CORD AND HINDBRAIN OF THE CHICK

Because the microinjection technique involves no species-specific reagents, it has been used to study the fate of cells in several systems, including the developing avian neural tube. In the spinal cord, clones were dispersed over considerable distances, spreading rostrocaudally as much as a somite length. No significant boundaries to this spreading were seen; in a few cases, apparent limitations were due to physical interactions with the somites (Stern *et al.*, 1991). Individual precursors in the developing spinal cord gave rise to multiple cell types, in agreement with studies using retroviral tracers in birds (Leber *et al.*, 1990) and using injected horseradish peroxidase in amphibians (Hartenstein, 1989). These results indicated that the spinal cord, like the neural retina, is derived from multipotent precursors.

In contrast to other axial levels, single precursors in the rhombencephalon gave rise to clones that were restricted, both in position and phenotype. The clones in the hindbrain were relatively large, containing 8 to 32 cells, and intermixed extensively with descendants from neighboring neuroblasts. Despite this dramatic intermixing, the descendants failed to cross the segment boundaries in almost all cases, suggesting that cell movement was restricted by the boundaries between the rhombomeres (hindbrain segments; see Keynes and Lumsden, 1990). To test the timing and placement of these restrictions, the injections were performed at rhombomere boundaries both before and after overt rhombomere appearance. If the precursors were labeled after the boundaries became visible, the descendants were always restricted to one rhombomere; if the precursors were labeled before the boundaries appeared, a few of the clones had descendants in two adjacent rhombomeres (Fraser *et al.*, 1990). These results suggest that a barrier to cell mixing appears at the time of rhombomere formation.

The precursors in the hindbrain were restricted also in the formation of cell phenotypes. Most of the clones contained only one neuronal phenotype, while fewer than 15% of the clones showed two distinct phenotypes (A. Lumsden, S. E. Fraser, and R. Keynes, in preparation). These unipotent clones were extensively intermixed with the descendants of neighboring cells that were themselves developing into other neuronal phenotypes. The relatively large clone sizes and the intermixing of clones argues against these results being the product of either chance or extrinsic determinants that were spatially restricted. In other words, the unipotency of these clones suggests that some of the precursor cells in the avian rhombencephalon are committed to a specific phenotype well before the last cell divisions of the lineage. Of course, such data can be only suggestive; definitive conclusions concerning the commitment of a precursor require an analysis of its fate following an experimental perturbation.

CONCLUDING REMARKS

Recent cell lineage studies have indicated that the vertebrate nervous system uses a wide range of developmental strategies for specifying cell fate. One extreme strategy is the specification of the final cell fate at or after the last cell division. This mode is suggested by the multipotent precursors that have been described in the neural retina (chick: Fekete *et al.*, 1990; rodent: Turner and Cepko, 1987; Turner *et al.*, 1990; frog: Holt *et al.*, 1988; Wetts and Fraser, 1988; Wetts *et al.*, 1989), the spinal cord (frog: Hartenstein, 1989; chick: Leber *et al.*, 1990; Stern *et al.*, 1988), the optic tectum (chick: Galileo *et al.*, 1990; Gray *et al.*, 1988; Gray and Sanes, 1991), and the neural crest (chick: Bronner-Fraser and Fraser, 1988, 1989). In contrast, cells in other regions of the nervous system apparently make decisions about cell fate before the end of the lineage. Clones in the rhombencephalon show both spatial and phenotypic restrictions; precursors typically give rise to single-phenotype clones that do not cross segment boundaries. Microinjection studies of the avian diencephalon revealed similar cell mixing restrictions at the neuromere boundaries (M. Figdor and C. Stern, personal communication); phenotypic restrictions have not yet been established. It appears that the mammalian cerebral cortex uses an intermediate strategy: some precursors are multipotent (Price and Thurlow, 1988; Walsh and Cepko, 1988), but the precursors appear to become restricted in potential near the end of the lineage (Barfield *et al.*, 1990; Luskin *et al.*, 1988; McConnell and Kaznowski,

1991; Parnavelas *et al.*, 1990). These regional differences in the developmental potential of the neural precursors provide opportunities to perform comparative analyses that will yield unique insights into the mechanisms underlying the specification of cell phenotype, position, and number. Because the microinjection technique can be used in any region of the nervous system, it is well suited to a comparative approach on different regions and species.

ACKNOWLEDGMENTS

This work was supported by grants from the NIH (EY08153, EY08363, and HD25390) and a gift from the Monsanto Corp. We thank M. S. Carhart, S. Burgan, and T. Joe for their technical assistance.

REFERENCES

Adler, R., and Farber, D., eds. (1986). "The Retina: A Model for Cell Biology Studies," part I and II. Academic Press, Orlando.

Austin, C. P., and Cepko, C. L. (1990). Cellular migration patterns in the developing mouse cerebral cortex. *Development* **110**, 713–732.

Barfield, J. A., Parnavelas, J. G., and Luskin, M. B. (1990). Separate progenitor cells give rise to neurons, astrocytes, and oligodendrocytes in the rat cerebral cortex. *Soc. Neurosci. Abstr.* **16**, 1272.

Beach, D. H., and Jacobson, M. (1979). Patterns of cell proliferation in the retina of the clawed frog during development. *J. Comp. Neurol.* **183**, 603–614.

Bronner-Fraser, M., and Fraser, S. E. (1988). Cell lineage analysis reveals multipotency of some avian neural crest cells. *Nature* **335**, 161–164.

Bronner-Fraser, M., and Fraser, S. E. (1989). Developmental potential of avian trunk neural crest cells *in situ. Neuron* **3**, 755–766.

Dowling, J. E. (1976). Physiology and morphology of the retina. *In* "Frog Neurobiology" (R. Llinas and W. Precht, eds), pp. 278–296. Springer-Verlag, New York.

Dowling, J. E. (1987). "The Retina: An Approachable Part of the Brain." Belknap Press, Cambridge, MA.

Fekete, D. M., Ryder, E. F., Stoker, A. W., and Cepko, C. L. (1990). Neuronal lineage and determination in the chick retina using retroviruses and cell transplants. *Soc. Neurosci. Abstr.* **16**, 1272.

Fraser, S. E., Keynes, R., and Lumsden, A. (1990). Segmentation in the chick embryo hindbrain is defined by cell lineage restrictions. *Nature* **344**, 431–435.

Galileo, D. S., Gray, G. E., Owens, G. C., Majors, J., and Sanes, J. R. (1990). Neurons and glia arise from a common progenitor in chicken optic tectum: demonstration with two retroviruses and cell type-specific antibodies. *Proc. Natl. Acad. Sci. U.S.A.* **87**, 458–462.

Gimlich, R. L., and Braun, J. (1985). Improved fluorescent compounds for tracing cell lineage. *Dev. Biol.* **109**, 509–514.

Gray, G. E., and Sanes, J. R. (1991). Migratory paths and phenotypic choices of clonally related cells in the avian optic tectum. *Neuron* **6**, 211–225.

Gray, G. E., Glover, J. C., Majors, J., and Sanes, J. R. (1988). Radial arrangement of clonally related cells in the chicken optic tectum: Lineage analysis with a recombinant retroviruses. *Proc. Natl. Acad. Sci. U.S.A.* **85**, 7356–7360.

Hartenstein, V. (1989). Early neurogenesis in *Xenopus:* The spatio-temporal pattern of proliferation and cell lineages in the embryonic spinal cord. *Neuron* **3**, 399–411.

Holt, C. E., Bertsch, T. W., Ellis, H. M., and Harris, W. A. (1988). Cellular determination in the *Xenopus* retina is independent of lineage and birth date. *Neuron* **1**, 15–26.

Jacobson, M. (1968). Cessation of DNA synthesis in retinal ganglion cells correlated with the time of specification of their central connections. *Dev. Biol.* **17**, 219–232.

Keynes, R., and Lumsden, A. (1990). Segmentation and the origin of regional diversity in the vertebrate central nervous system. *Neuron* **2**, 1–9.

Kimmel, C. B., and Warga, R. M. (1986). Tissue-specific cell lineages originate in the gastrula of the zebrafish. *Science* **231**, 365–368.

Leber, S. M., Breedlove, S. M., and Sanes, J. R. (1990). Lineage, arrangement, and death of clonally related motoneurons in chick spinal cord. *J. Neurosci.* **10**, 2451–2462.

Luskin, M. B., Pearlman, A. L., and Sanes, J. R. (1988). Cell lineage in the cerebral cortex of the mouse studied *in vivo* and *in vitro* with a recombinant retrovirus. *Neuron* **1**, 635–647.

McConnell, S. K., and Kaznowski, C. E. (1991). Cell cycle dependence of laminar determination in developing neocortex. *Science* **254**, 282–285.

Nieuwkoop, P. D., and Faber, J. (1956). "Normal Table of *Xenopus laevis* (Daudin)." Elsevier–North-Holland, Amsterdam.

Parnavelas, J. G., Barfield, J. A., and Luskin, M. B. (1990). Lineage relationships of pyramidal and nonpyramidal neurons in the rat cerebral cortex. *Soc. Neurosci. Abstr.* **16**, 1272.

Price, J., and Thurlow, L. (1988). Cell lineage in the rat cerebral cortex: A study using retroviral-mediated gene transfer. *Development* **104**, 473–482.

Ramon y Cajal, S. (1972). "The Structure of the Retina" (compiled and transl. by S. A. Thorpe and M. Glickstein). Thomas, Springfield, IL.

Sanes, J. R. (1989). Analysing cell lineage with a recombinant retrovirus. *Trends Neurosci.* **12**, 21–28.

Sheard, P., and Jacobson, M. (1987). Clonal restriction boundaries in *Xenopus* embryos shown with two intracellular lineage tracers. *Science* **236**, 851–854.

Stern, C. D., Fraser, S. E., Keynes, R. J., and Primmett, D. R. (1988). A cell lineage analysis of segmentation in the chick embryo. *Development* **104**(Suppl), 231–244.

Stern, C. D., Jaques, K. F., Lim, T.-M., Fraser, S. E., and Keynes, R. J. (1991). Segmental lineage restrictions in the chick embryo spinal cord depend on the adjacent somites. *Development* **113**, 239–244.

Turner, D. L., and Cepko, C. L. (1987). A common progenitor for neurons and glia persists in rat retina late in development. *Nature* **328**, 131–136.

Turner, D. L., Snyder, E. Y., and Cepko, C. L. (1990). Lineage-independent determination of cell type in the embryonic mouse retina. *Neuron* **4**, 833–845.

Walsh, C., and Cepko, C. L. (1988). Clonally related cortical cells show several migration patterns. *Science* **241**, 1342–1345.

Warga, R. M., and Kimmel, C. B. (1990). Cell movements during epiboly and gastrulation in zebrafish. *Development* **108**, 569–580.

Wetts, R., and Fraser, S. E. (1988). Multipotent precursors can give rise to all major cell types of the frog retina. *Science* **239,** 1142–1145.

Wetts, R., and Fraser, S. E. (1989). Slow intermixing of cells during *Xenopus* embryogenesis contributes to the consistency of the blastomere fate map. *Development* **105,** 9–15.

Wetts, R., Serbedzija, G. N., and Fraser, S. E. (1989). Cell lineage analysis reveals multipotent precursors in the ciliary margin of the frog retina. *Dev. Biol.* **136,** 254–263.

6

Axon Guidance in the Mammalian Spinal Cord

JANE DODD* AND THOMAS M. JESSELL†

*Departments of *Physiology and Cellular Biophysics and †Biochemistry
and Molecular Biophysics, Columbia University
New York, New York 10032*

INTRODUCTION

The diverse functions of the nervous system, from simple reflex responses to cognition, depend on the ordered interconnections of hundreds of morphologically distinct classes of neurons. The generation of patterned neural networks begins with the extension of axons toward their targets and the formation of selective synaptic contacts (see Ref. 1). Some of the cellular mechanisms that contribute to the formation of neuronal circuits during vertebrate development, in particular the mechanisms contributing to the patterning of axonal projections in the developing mammalian spinal cord, are discussed here.

The histological analysis of individual embryonic neurons by Ramon y Cajal (2) introduced the possibility that axons select defined pathways and are guided to their targets. This idea was reinforced by the observations of stereotyped growth of axons that project to their targets along complex but highly reproducible pathways in living amphibian embryos (3). These experiments, coupled with Sperry's proposals that the selectivity of connections formed during early development depends on the recognition of specific chemical markers present on individual cell types, have provided the conceptual framework for most subsequent studies of axon guidance.

With the availability of cellular assays of axonal guidance and growth cone recognition for the analysis of developing systems, it has become increasingly clear that the growth and guidance of embryonic axons is dependent on molecular cues in their environment. The first compelling evidence for specificity in the pathfinding of de-

81

*Cell–Cell Signaling in
Vertebrate Development*

veloping axons derived from studies of motor axon pathfinding to muscle targets in the chick embryo (5) and from the analysis of growth cone recognition in insects (6,7).

Several distinct types of cue or mechanisms of guidance have been implicated in axon guidance. Local variations in the adhesive properties of epithelial or neural cells and in the expression of extracellular matrix components on which axons are initially extended, and over which they subsequently migrate, may provide permissive or inhibitory substrate cues that influence pathway choice (8–10). Distinct pathways mediated by axonal glycoproteins may also be formed by the migration of growth cones along subsets of axon fascicles (11). In addition to contact-based cues, growth cones can orient in response to gradients of diffusible factors that emanate from restricted populations of target cells (12–14).

An axon potentially encounters combinations of these and many other guidance cues sequentially and, in effect, projects to its final target in a series of short-range extensions under the influence of spatially restricted information. To achieve this, growth cones and axons may have to adapt to their changing environment (see Ref. 1). Marked alterations in the morphology of growth cones have been observed during the migration of axons through different cellular environments (15–17) suggesting that growth cones actively respond to the environment and seek out guidance cues. In addition, temporally and spatially regulated alterations in the expression of axonal surface proteins that may mediate cellular interactions have been observed (1,18,19). The final patterning of neuronal connections thus appears to result from coordinated interactions between growth cones and their cellular environment.

In an attempt to define more clearly the initial steps in the formation of neural circuits in the mammalian central nervous system we have examined the properties and trajectory of a class of commissural neurons in the embryonic rodent spinal cord. We have focused on the spinal cord as a model system because it represents one of the simplest and most conserved regions of the central nervous system (CNS). In addition, the physiological analysis of spinal cord circuitry (20,21) has provided a basic framework for the understanding of sensory and motor function. Studies of spinal cord development may eventually provide new insights into the strategies used to control the formation and function of neural circuits in the central nervous system.

Commissural neurons differentiate in the dorsal region of the rat spinal cord (18,22,23) and can be identified early in development by the selective expression of the 135 kDa glycoprotein, TAG-1 (18,24). The earliest projecting commissural neurons extend axons over a neuroepithelium that is relatively free of other axons and follow a complex course during which they appear to make several navigational choices. Commissural axons project ventrally, initially parallel with and close to the external limiting membrane of the neural tube (Fig. 1). Just dorsal to the motoneuron pool, the growth cones alter course and extend medially toward the ventral midline where they cross a group of specialized epithelial cells, the floor plate, before turning rostrally to join the contralateral ventral funiculus. The re-

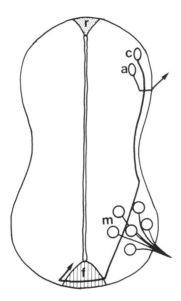

Fig. 1. Axonal trajectories of early differentiating neurons in the embryonic rat spinal cord. A schematic diagram of a transverse section of an E11–12 rat spinal cord shows the location of the first three classes of differentiated neurons and their prospective axonal trajectories. Motoneurons (m) differentiate in the ventral region of the spinal cord and extend out of the spinal cord axons to their target muscles. Commissural (c) and association (a) neurons differentiate in the dorsal region of the spinal cord, adjacent to the roof plate (r). Association axons project laterally to join the ipsilateral lateral funiculus. Commissural axons grow ventrally along the lateral margin of the spinal cord to the motor column, then alter their trajectory and course directly through the nascent motor column to the floor plate (f). These axons appear to project to the floor plate through a cellular environment that does not contain a performed substrate pathway. After crossing the midline of the spinal cord at the floor plate, commissural axons turn by 90° to form longitudinal projections in the contralateral ventrolateral funiculus.

sults of experiments in rat, chick, and zebrafish suggest that the floor plate plays a prominent part in commissural axon guidance, acting as an intermediate target in their trajectory. At later stages of development, commissural axons also take the highly stereotyped route described above through the spinal cord but may do so by interacting with the axons of earlier projecting axons.

ORIENTATION OF COMMISSURAL AXONS IN RESPONSE TO A FLOOR PLATE-DERIVED CHEMOATTRACTANT

The directed growth of commissural axons in the chick spinal cord, ventromedially through the motoneuron pool and toward the floor plate, prompted Ramon y Cajal (25) to suggest that diffusible cues might guide commissural axons over this domain. Chemotaxis is a commonly used mechanism of cell orientation in nonneuronal systems (26) and there is increasing evidence that growth cones can orient in response to gradients of chemoattractant molecules that are released selectively by cells that form the intermediate or final targets of a developing axon. An early candidate for a chemoattractant acting in the peripheral nervous system (PNS) and CNS was nerve growth factor (NGF). *In vivo,* injection of high concentrations of NGF into the CNS results in profuse abnormal growth of sympathetic axons toward the injection site (27). *In vitro,* sensory neurons reorient growth cones toward a point source of NGF (28,29) by a mechanism that appears to be independent of the trophic effect of NGF on these neurons. Sympathetic and sensory neurons have also been shown to reorient up a gradient of NGF in three-dimensional matrices (30,31). More recent experimental evidence suggests that diffusible factors other than NGF may be involved in the development of selective connections in both the peripheral and the central nervous systems. Mouse trigeminal ganglion (TGG) neurons isolated *in vitro* have been shown to grow selectively toward a potential target tissue, the epithelium of the maxillary arch, placed at a distance from the TGG *in vitro* (12,32). Similar studies have suggested that the axons of corticospinal projection neurons extend collateral branches into one of their final targets, the basilar pons, in response to a diffusible factor secreted by the pontine tissue (33).

In intact embryos, it is difficult to infer from the trajectory of an axon whether it is guided by cues provided by cells along the substrate pathway or by chemotropic factors secreted by distant cellular targets. Evidence for chemotropic guidance in the nervous system has therefore relied primarily on *in vitro* assays (34). The involvement of the floor plate in the directed growth of commissural axons has been tested by assaying the effect of the floor plate on commissural axon outgrowth from embryonic dorsal spinal cord explants *in vitro* (13,14). When cultured alone, little or no axon outgrowth occurs from dorsal explants (Fig. 2a). In contrast, extensive axon outgrowth occurs when dorsal explants are cultured with a floor plate explant placed at a distance of 100–400 μm (Fig. 2b). Under these conditions, most axons project in thick fascicles from the ventralmost edge of the explant and grow toward the floor plate. In contrast, when dorsal explants are cultured alone but exposed to medium conditioned by the floor plate, extensive axon outgrowth occurs from all edges of the explant (14). The axons that project from a dorsal explant in the presence of a floor plate explant express TAG-1, suggesting that they derive from commissural neurons. Moreover, the effect of the floor plate on axon outgrowth appears to be selective for commissural neu-

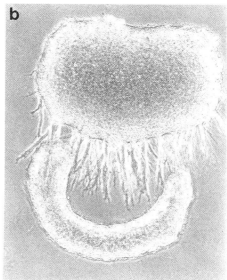
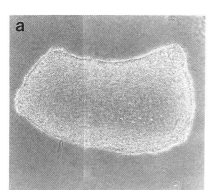

Fig. 2. The floor plate promotes commissural axon outgrowth from dorsal spinal cord explants. An explant of the dorsal third of an E11 spinal cord was cultured in a three-dimensional collagen matrix for 40 hr alone (a) or with its ventralmost edge facing an explant of the floor plate (b). Extensive axon outgrowth occurs only when the floor plate explant is present.

rons. Axons that do not express TAG-1 and that presumably derive from other sensory relay neurons present within dorsal explants do not project from the explants. These experiments provide evidence that the floor plate secretes a diffusible factor that promotes commissural axon outgrowth from dorsal spinal cord explants but they do not reveal whether the floor plate can orient the growth of commissural axons.

To determine whether axons orient toward a local source of chemoattractant, the floor plate was examined for its ability to cause commissural axons to deviate from their initial dorsoventral (D–V) trajectory within a dorsal spinal cord explant (14). Commissural neurons located at distances greater than 300 μm from the floor plate failed to reorient their axons and continued to project along their normal D–V trajectory. In contrast, essentially all axons located within 300 μm of the floor plate deviated from their D–V trajectory and reoriented toward the floor plate. The floor plate therefore appears able to override intrinsic D–V polarity cues present within the dorsal neuroepithelium and to cause the reorientation of commissural axons within a neuroepithelial environment that is close to that encountered in the developing embryo. The distance over which axons reorient in these experiments could represent the maximum range of action of the chemoattractant or indicate simply the rate of diffusion of the factor. In the embryonic spinal cord, commissural axons break away from their initial ventral trajectory and grow toward the floor plate at a distance of about 100–200 μm from the ventral midline, well within the range of action of the chemoattractant factor defined *in vitro*. This is also within the theoretical limit of action of diffusible signals in embryonic tissues and consistent with the range over which diffusible factors have been proposed to influence cell pattern in other developing organisms (36,37). The sensitivity of growth cones in detecting diffusible gradients has not yet been determined. However, other vertebrate cells, for example, leukocytes, can orient in response to soluble gradients of chemotactic factors that generate only 1% differences in concentration across the diameter of the cell (38). It seems likely that growth cones are similarly sensitive in their ability to recognize cell-surface and diffusible guidance cues (39).

Identification of the chemoattractant factor(s) will clearly be necessary to define the contribution of chemotropism to the guidance of commissural axons *in vivo*. However, transplantation experiments provide preliminary evidence that the floor plate is capable of orienting commissural axons *in vivo* as well as *in vitro*. Grafting of small segments of floor plate adjacent to the neural tube in chick embryos resulted in the deviation of commissural axons from their normal ventral trajectory. At a point dorsal to the motor column the axons emerged from the neural tube and extended toward the ectopic floor plate (35). Commissural axons in the contralateral half of the spinal cord follow their usual path and project medially through the motor column toward the floor plate. Grafts of other regions of the neural tube do not result in the rerouting of commissural axons, suggesting that this effect is specific to the floor plate. Thus, in an extreme scenario, the presence and graded distribution of the factor(s) might be an absolute requirement for the maintained ventral growth of commissural axons after they have reached the motor column. Aternatively, commissural axons may be capable of growing ventrally beyond the motor column even in the absence of floor plate-derived factor. The role of the factor might then be to provide a cue that ensures that commissural axons project directly through the motor column to the ventral midline, rather than fasciculate with motor axons that project out of the spinal cord. The floor plate itself extends throughout the spinal cord, hindbrain, and midbrain, raising the possibility that the axons of commissural neurons in other regions of the CNS respond to the chemoattractant. Our findings, coupled with many examples of apparently directed axonal growth within the central nervous system in the absence of any other known cues, raise the possibility that chemotropism is a general mechanism of axon guidance within the developing central nervous system.

PATTERNING OF AXONS BY CONTACT-MEDIATED CUES AT THE FLOOR PLATE

The floor plate appears to have several roles in the guidance of commissural axons. Evidence from experiments with rat (40), zebrafish (41), and chick (42) embryonic spinal cord suggests that in addition to providing a diffusible chemoattractant, the floor plate guides commissural growth cones by contact-mediated cues. On arrival at the ventral midline in the rat, commissural growth cones cross the floor plate and immediately change trajectory, projecting rostrally in the contralateral ventral funiculus (Fig. 1). The bilateral symmetry of this projection raises the possibility that cues directing commissural growth cones into the ventral funiculus are present on both sides of the floor plate but can be recognized only after growth cones have traversed the midline and undergone some modification.

This idea has received support from the finding that as the growth cones of commissural axons cross the floor plate, expression of TAG-1 by the entire commissural neuron ceases (Figs. 3a and 3c). At the same time, the distal contralateral axonal (Figs. 3b and 3d) segments of commissural axons initiate expression of the axonal glycoprotein, L1 (18). Coincident with the change in glycoprotein expression on distal portions of the commissural axons, the relationship of the axons relative to each other changes. The commissural axons alter their trajectory from an apparently nonfasciculated projection in the transverse plane of the spinal cord to a fasciculated tract parallel with the longitudinal axis of the spinal cord. Passage across the floor plate epithelium therefore appears to have a profound effect on the

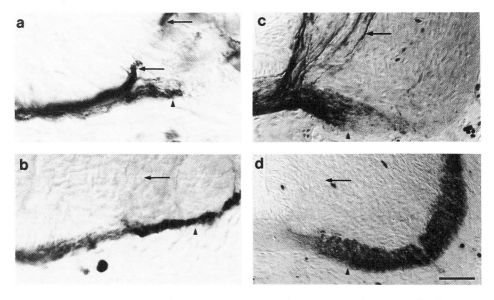

Fig. 3. Transitions in glycoprotein expression by commissural axons at the floor plate. (a,c) In E12 (a) and E14 (c) rat spinal cord the TAG-1 glycoprotein, visualized by immunoperoxidase staining, is expressed on commissural axons (arrows) as they project toward the floor plate. As commissural axons join the ventral funiculus, indicated by arrowheads, expression of TAG-1 ceases. (b,d) L1 is detected only at very low levels on the proximal segment of commissural axons during their extension toward the floor plate (arrows). L1 is expressed at high levels on the contralateral segment of commissural axons that have joined the ventral funiculus. This transition in axonal glycoprotein expression occurs at the floor plate. Bar: (a,b) 20 μm; (c,d) 50 μm. From Dodd *et al.* (18).

properties of commissural neurons that may be responsibie for their navigation into the contralateral ventral funiculus.

To begin to determine whether the floor plate functions *in vivo* to provide local guidance cues we have examined the morphology and trajectory of commissural growth cones as they navigate the floor plate. The axons and growth cones of commissural neurons have been labeled in a semi-intact preparation of embryonic spinal cord with the fluorescent carbocyanine dye, DiI (1,1'-dioctadecyl-3,3,3', 3'-tetramethylindo-carbocyanine perchlorate) (40). The projection pattern of individual axons is precise and stereotyped. A striking feature of the trajectory of these axons is their contrasting behavior on arrival at the two edges of the floor plate: the first axons to arrive at the ipsilateral edge continue straight across the edge without altering their course. At the contralateral edge the rostral turn is abrupt and at an angle of approximately 90° (Fig. 4). The axons appear then to maintain contact with the longitudinal face of the floor plate for a further 30–100 μm. Virtually all axons display this stereotyped course. Later arriving growth cones also make orthogonal turns at the contralateral face. After having turned, these axons only then appear to fasciculate on the established, earlier arriving axons. This suggests that all the growth cones turn rostrally in response to cues associated with the underlying floor plate cells or neuroepithelium. However, ultrastructural studies are necessary to determine whether axons turn completely independently of each other.

Studies of developing invertebrate axonal projections have provided evidence for the existence of intermediate cellular targets, known as landmark or guidepost cells, that influence axonal trajectories, providing contact-mediated guidance cues for growth cones (6,43,44). While the existence of cells that serve equivalent functions in vertebrates has not been established, floor plate cells in the vertebrate CNS may have properties stikingly similar to those of invertebrate guidepost cells. The role of guidepost cells in axon guidance in invertebrates has been inferred from the axonal misrouting that results after selective ablation of the target cells and in genetic mutants in which target cells fail to differentiate (6,45,46). To examine further the role of the floor plate in guidance at the midline we analyzed commissural axon trajectories in the mouse mutant, Danforth's short-tail (*Sd*), in which the floor plate is absent in caudal regions of the neuraxis (47,48). In the affected region of the spinal cord, commissural axons exhibit aberrant trajectories as they reach and cross the ventral midline (49), supporting the idea that the floor plate provides midline guidance cues and emphasizing the similarities that are

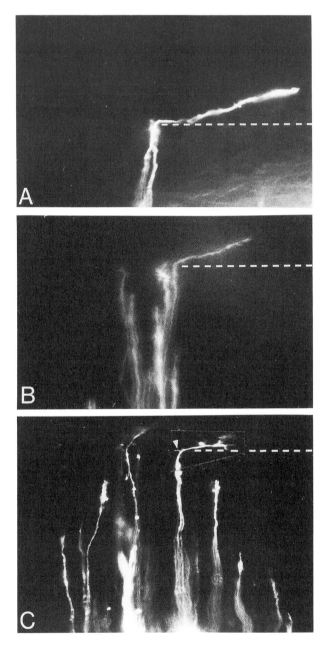

Fig. 4. Commissural growth cones make orthogonal turns at the contralateral edge of the floor plate. Fluorescence micrographs of three whole-mount preparations of E12.5–13 rat spinal cord. Commissural axons and growth cones have been labeled with DiI. In each case the contralateral edge of the floor plate, discernible under phase optics, is indicated by dotted lines. Rostral direction is to the right. In (A) after the

emerging between mechanisms of axon guidance in different organisms.

One difficulty in interpreting these results in that the *Sd* mutation also results in a decrease in the number of motoneurons in the spinal cord, consistent with the results of experiments in chick embryos described above. A perturbation of the organization of other cell types in the ventral spinal cord could, then, contribute to the disruption in commissural axon pathfinding. However, studies in other vertebrate species have provided evidence that the pattern of commissural axon growth is perturbed in the absence of a floor plate under conditions in which the remaining cell types in the neural tube are less likely to be affected (50,51). Thus it is likely that the floor plate contributes directly to the guidance of commissural axons.

The morphology of commissural growth cones changes dramatically as they cross the floor plate. Throughout their course toward the floor plate the growth cones have an elongated and simple shape. At the ipsilateral edge of the floor plate they become expanded and complex with several long filopodia. This complexity is maintained as the growth cones turn to extend into the longitudinal plane (40). Growth cone complexity has been thought to reflect the exploratory response of a nerve ending to a changing environment (15–17) and may be an indication in commissural neurons of adhesive guidance cues present at the midline. The floor plate expresses several surface proteins that appear to promote cell adhesion or neurite outgrowth and these may mediate interactions between floor plate cells and commissural axons. For example, both the floor plate and the commissural axons express high levels of the neural cell adhesion molecule, NCAM (18,49). The neurite outgrowth-promoting molecule P84 (52) is expressed selectively by floor plate cells as commissural axons cross the midline. In addition, the floor plate expresses high levels of a novel outgrowth-promoting molecule, F-spondin (53). The basal lamina that underlies the floor plate is also antigenetically distinct (45) raising the possibility that changes in axonal trajectory at the floor plate involve growth cone recognition of specialized basal lamina components (42,49). Midline cells of the floor plate in the spinal

turn, the axon follows the floor plate–neuroepithelial boundary initially and then moves laterally. In (B) and (C) the preparation has been flattened slighty to reveal the longitudinal face of the floor plate, along which the axons have traveled. (A,B) × 350; (C) × 250. From Bovolenta and Dodd (40).

cord and more rostrally at the optic chiasm represent passageways for axons crossing from one side of the brain to the other. At both the floor plate and the optic chiasm crossing axons undergo substantial reorganization of trajectory (54,55), suggesting that these cells play a major role in modulating the bilateral organization of crossed pathways.

SUMMARY

Our findings suggest that one class of axons employs a variety of navigational cues and that the order of presentation and relative strengths of these cues are crucial to the establishment of the final pathway. In addition, although the superficial plans of vertebrate and invertebrate nervous systems are quite distinct, specialized midline cell groups appear to have a critical role in establishing neuronal projection patterns in animals as divergent as insects and mammals. Studies of axonal guidance in *Drosophila* have recently identified midline glial cells which appear to have important roles in commissural axon pathfinding (19,56). Molecular cloning has also revealed that many proteins implicated in cell adhesion and axon fasciculation in vertebrates and invertebrates share structural features (19,57). Moreover, invertebrate axonal glycoproteins such as fasciclin I and fasciclin II exhibit striking spatial restrictions at the ventral midline in a manner similar to that described in the spinal cord for TAG-1 and L1. The genetic analysis of axonal pathfinding in the nematode worm, *C. elegans*, has also defined genes responsible for distinct D–V and anteroposterior (A–P) domains in the growth of commissural axons (58,59) supporting the idea that distinct guidance cues operate at different stages in the trajectory of a single axon.

The convergence of ideas from these diverse embryonic systems suggests that it may be possible to define a set of general principles and molecular mechanisms that underlie the control of neural cell pattern and the formation of neuronal connections in different organisms.

ACKNOWLEDGMENTS

We thank our many colleagues who have contributed practically and theoretically to this work. The work was funded by grants to J.D. from the National Science Foundation, The Irma T. Hirschl Fund, and The Klingenstein Foundation. T.M.J. is an Investigator of the Howard Hughes Medical Institute.

REFERENCES

1. Dodd, J., and Jessell, T. M. (1988). Axon guidance and the patterning of neuronal projections in vertebrates. *Science* **242**, 692–699.
2. Ramon Y Cajal, S. (1909). *In* "Histologie du Systeme Nerveux de l'Homme et des Vertebres," Vol. 1, pp. 657–664. Consejo Superior de Investigaciones Cientificas, Madrid.
3. Spiedel, C. C. (1941). Adjustments of nerve endings. *Harvey Lect.* **36**, 126–158.
4. Sperry, R. W. (1963). Chemoaffinity in the orderly growth of nerve fiber patterns and connections. *Proc. Natl. Acad. Sci. U.S.A.* **50**, 703–710.
5. Lance-Jones, C., and Landmesser, L. (1981). Pathway selection by chick lumbo-sacral motoneurons during normal development. *Proc. R. Soc. Lond. (Biol.)* **214**, 1–18.
6. Bentley, D., and Keshishian, H. (1982). Pathfinding by peripheral pioneer neurons in grasshoppers. *Science* **218**, 1082–1088.
7. Goodman, C. S., Bastiani, M. J., Doe, C. Q., du Lac, S., Helfand, S. L., Kuwada, J. Y., and Thomas, J. B. (1984). Cell recognition during neuronal development. *Science* **225**, 1271–1279.
8. Kapfhammer, J. P., Grunewald, B. E., and Raper, J. A. (1986). The selective inhibition of growth cone extension by specific neurites in culture. *J. Neurosci.* **6**, 2527–2534.
9. Stern, C. D., Sisodiya, S. M., and Keynes, R. J. (1986). Interactions between neurites and somite cells: Inhibition and stimulation of nerve growth in the chick embryo. *J. Embryol. Exp. Morphol.* **91**, 209–226.
10. Walter, J., Henke-Fahle, S., and Bonhoeffer, F. (1987). Avoidance of posterior tectal membranes by temporal retinal axons. *Development* **101**, 909–913.
11. Rathjen, F. G. (1988). A neurite outgrowth-promoting molecule in developing fiber tracts. *Trends Neurosci.* **11**, 183–184.
12. Lumsden, A. G. S., and Davies, A. (1986). Chemotropic effect of specific target epithelium in development of the mammalian nervous system. *Nature* **323**, 538–539.
13. Tessier-Lavigne, M., Placzek, M., Lumsden, A. G. S., Dodd, J., and Jessell, T. M. (1988). Chemotropic guidance of developing axons in the mammalian central nervous system. *Nature* **336**, 775–778.
14. Placzek, M., Tessier-Lavigne, M., Jessell, T. M., and Dodd, J. (1990). Orientation of commissural axons in vitro in response to a floor plate derived chemoattractant. *Development* **110**, 19–30.
15. Tosney, K. W., and Landmesser, L. (1985). Growth cone morphology and trajectory in the lumbosacral region of the chick embryo. *J. Neurosci.* **5**, 2345–2358.
16. Bovolenta, P., and Mason, C. (1987). Growth cone morphology varies with position in the developing mouse visual pathway from retina to first targets. *J. Neurosci.* **7**, 1447–1460.
17. Raper, J. A., Bastiani, M. J., and Goodman, C. S. (1983). Pathfinding by neuronal growth cones in grasshopper embryos. I. Divergent choices made by the growth cones of sibling neurons. *J. Neurosci.* **3**, 20–30.
18. Dodd, J., Morton, S. B., Karagogeos, D., Yamamoto, M., and Jessell, T. M. (1988). Spatial regulation of axonal glycoprotein expression on subsets of embryonic spinal neurons. *Neuron* **1**, 105–116.
19. Harrelson, A. L., and Goodman, C, S. (1988). Growth cone guidance in insects: Fasciclin II is a member of the immunoglobulin superfamily. *Science* **242**, 700–708.

20. Sherrington, C. S. (1906). In "The Integrative Action of the Nervous System," Yale Univ. Press, New Haven, 2nd Ed., 1947.

21. Eccles, J. C. (1964). In "The Physiology of Synapses." Springer, New York.

22. Holley, J. (1982). Early development of the circumferential axonal pathway in mouse and chick spinal cord. *J. Comp. Neurol.* **205**, 371–382.

23. Wentworth, L. (1984). The development of the cervical spinal cord of the mouse embryo II A golgi analysis of sensory, commissural and association cell differentiation. *J. Comp. Neurol.* **222**, 96–115.

24. Furley, A. J., Morton, S. B., Manalo, D., Karagogeos, D., Dodd, J., and Jessell, T. M. (1990). The axonal glycoprotein TAG-1 is an immunoglobulin superfamily member with neurite outgrowth promoting activity. *Cell* **61**, 157–170.

25. Ramon Y Cajal, S. (1892). Le retine des vertebres. *La Cellule* **9**, 121–246. In "The Structure of the Retina" (S. A. Thorpe and M. Glickstein (comp. and transl)). Thomas, Springfield, IL., 1972.

26. Trinkaus, J. P. (1985). Further thoughts on directional cell movement during morphogenesis. *J. Neurosci. Res.* **13**, 1–19.

27. Menesini-Chen, M. G., Chen, J. S., and Levi-Montalcini, R. (1978). Sympathetic nerve fiber ingrowth in the central nervous system of neonatal rodents upon intracerebral NGF injection. *Arch. Ital. Biol.* **116**, 53–84.

28. Gundersen, R. W., and Barrett, J. N. (1979). Neuronal chemotaxis: Chick dorsal-root axons turn toward high concentrations of nerve growth factor. *Science* **206**, 1079–1080.

29. Gundersen, R. W., and Barrett, J. N. (1980). Characterization of the turning response of dorsal root neurites toward nerve growth factor. *J. Cell Biol.* **87**, 546–554.

30. Letourneau, P. C. (1978). Chemotactic response of nerve fiber elongation to nerve growth factor. *Dev. Biol.* **66**, 183–196.

31. Ebendal, T., and Jacobson, C. O. (1977). Test of possible role of NGF in neurite outgrowth stimulation exerted by glial cells and heart explants in culture. *Brain Res.* **131**, 373–378.

32. Lumsden, A. G. S., and Davies, A. (1983). Earliest sensory nerve fibres are guided to peripheral targets by attractants other than nerve growth factor. *Nature* **306**, 786–788.

33. Heffner, C., Lumsden, A. G. S., and O'Leary, D. (1990). Target control of collateral extension and directional axon growth in the mammalian brain. *Science* **247**, 217–220.

34. Tessier-Lavigne, M., and Placzek, M. (1991). Target attraction: Are developing axons guided by chemotropism? *Trends Neurosci.* **14**, 303–310.

35. Wolpert, L. (1969). Positional information and the spatial pattern of cellular differentiation. *J. Theor. Biol.* **25**, 1–49.

36. Crick, F. H. C. (1970). Diffusion in embryogenesis. *Nature* **225**, 420–422.

37. Zigmond, S. H. (1977). The ability of polymorphonuclear leukocytes to orient gradients of chemotactic factors. *J. Cell Biol.* **75**, 606–616.

38. Walter, J., Allsop, T. E., and Bonhoeffer, F. (1990). A common denominator of growth cone guidance and collapse? *Trends Neurosci.* **13**, 447–452.

39. Placzek, M., Tessier-Lavigne, M., Yamada, T., Dodd, J., and Jessell, T. M. (1990). The guidance of developing axons by diffusible chemoattractants. *Cold Spring Harbor Symp.* **55**, 279–289.

40. Bovolenta, P., and Dodd, J. (1990). Guidance of commissural growth cones at the floor plate in the embryonic rat spinal cord. *Development* **109**, 435–447.

41. Kuwada, J. Y., Bernhardt, R. R., and Chitnis, A. B. (1990). Pathfinding by identified growth cones in the spinal cord of zebrafish embryos. *J. Neurosci.* **10,** 1229–1308.

42. Yaginuma, H., Homma, S., Kunzi, R., and Oppenheim, R. W. (1991). Pathfinding by growth cones of commissural interneurons in the chick embryo spinal cord: A light and electron microscopic study. *J. Comp. Neurol.* **304,** 78–102.

43. Bate, C. M. (1976). Pioneer neurons in an insect embryo. *Nature* **260,** 54–56.

44. Taghert, P., Bastiani, M. J., Ho, R. K., and Goodman, C. S. (1982). Guidance of pioneer growth cones: Filopodial contacts and coupling revealed with an antibody to Lucifer yellow. *Dev. Biol.* **94,** 391–399.

45. Bentley, D., and Caudy, M. (1983). Pioneer axons lose directed growth after selective killing of guidepost cells. *Nature* **304,** 62–64.

46. Thomas, J. B., Crews, S. T., and Goodman, C. S. (1985). Molecular genetics of the single-minded locus: a gene involved in the development of the *Drosophila* nervous system. *Cell* **52,** 133–141.

47. Gruneberg, H. (1958). Genetical studies on the skeleton of the mouse. XXII. The development of Danforth's short tail. *J. Embryol. Exp. Morphol.* **6,** 124–148.

48. Theiler, K. (1959). Anatomy and development of the "truncate" (boneless) mutation in the mouse. *Am. J. Anat.* **104,** 319–343.

49. Bovolenta, P., and Dodd, J. (1991). Perturbation of neuronal differentiation and axon guidance in the spinal cord of mouse embryos lacking a floor plate: Analysis of Danforth's short-tail mutation. *Development* **113,** 625–639.

50. Bernhardt, R. R., and Kuwada, J. Y. (1990). Floor plate ablations induce axonal pathfinding errors by spinal commissural cells in the zebrafish embryo. *Soc. Neurosci. Abstr.* **16,** 139.2.

51. Clarke, J. D. W., Holder, N., Soffe, S. R., and Storm-Mathisen, J. (1991). Neuroanatomical and functional analysis of neural tube formation in notochordless Xenopus embryos; laterality of the ventral spinal cord is lost. *Development* **112,** 499–516.

52. Chuang, W., and Lagenauer, C. F. (1990). Central nervous system antigen P84 can serve as a substrate for neurite outgrowth. *Dev. Biol.* **137,** 219–232.

53. Klar, A., Baldassare, M., and Jessell, T. M. (1992). F-spondin, a gene expressed at high level in the floor plate, encodes a secreted protein that promotes neural cell adhesion and neurite outgrowth. *Cell* **69,** 95–110.

54. Taylor, J. S. H. (1987). Fibre organization and reorganization in the retino-tectal projection of Xenopus. *Development* **99,** 393–410.

55. Godement, P., Salaun, J., and Mason, C. (1990). Retinal axon pathfinding in the optic chiasm: Divergence of crossed and uncrossed fibers. *Neuron* **5,** 173–186.

56. Klambt, C., Jacobs, J. R., and Goodman, C. S. (1991). The midline of the drosophila central nervous system: A model for the genetic analysis of cell fate, cell migration and growth cone guidance. *Cell* **64,** 801–815.

57. Jessell, T. M. (1988). Adhesion molecules and the hierarchy of neural development. *Neuron* **1,** 3–13.

58. Hedgecock, E. M., and Hall, D. H. (1990). Homologies in the neurogenesis of nematodes, arthropods and chordates. *Sem. Neurosci.* **2,** 159–172.

59. Hedgecock, E. M., Culotti, J. G., and Hall, D. H. (1990). The unc-5, unc-6, and unc-40 genes guide circumferential migrations of pioneer axons and mesodermal cells on the epidermis in C. elegans. *Neuron* **2,** 61–85.

7

Axon Patterning in the Visual System: Divergence of Retinal Axons to Each Side of the Brain at the Midline of the Optic Chiasm

CAROL A. MASON* AND PIERRE GODEMENT†
*Departments of Pathology
and Anatomy and Cell Biology
The Center for Neurobiology and Behavior
College of Physicians and Surgeons
Columbia University
New York, New York, 10032
and † Centre National de la Recherche Scientifique
Institute Alfred Fessard
Avenue de la Terrasse 91198 Gif-Sur-Yvette
France

A long-standing question in developmental biology is how populations of axons cross the midline of the young nervous system. The mouse visual system is an opportune model for axon guidance in this regard, since both a crossed and an uncrossed projection arise from each retina during development of the visual system of many vertebrates. While it has long been known that the site of divergence of these two populations occurs in the optic chiasm (24), the guidance mechanisms responsible for this bilateral projection are not well understood.

This otherwise well-studied system provides a good basis by which to analyze the problem of axon divergence. The topography of the

Cell–Cell Signaling in
Vertebrate Development

retinal projections and their functional characteristics are well known (e.g., 16, 42). The variety of ganglion cell types, their birthdates, and those of their corresponding targets have been established (16). Moreover, the origins of the subpopulations of retinal ganglion cells that project either ipsi- or contralaterally and their respective birthdates have been defined (10, 17, 53, 60). Finally, the relationships of retinal axon growth cones with neuronal and nonneuronal components in their outgrowth pathway, especially in the optic nerve, have been delineated (3, 4, 26).

Axon guidance has been predicated on a number of putative signaling mechanisms or cues. Studies on process outgrowth and growth cone navigation *in situ* in invertebrate and vertebrate model systems have implicated positive, growth-promoting cues in the substrates of pathways, cues expressed by nonneuronal cells or extracellular matrix, that mediate directional growth (e.g., 8, 14, 27, 54). Such a mechanism would allow for initial precision of pathway selection by axons (35). Crossing systems, such as the commissural neuron pathway in the spinal cord, rely on permissive cues provided by a midline structure, the floor plate (5, 15). One set of cues is dependent on growth cone substrate interactions and another on diffusible cues released by the midline structure (see Dodd and Jessell, Chapter 6).

In addition to growth-supporting mechanisms, inhibitory interactions are considered to play an important role (50), both in paths (47) and in targets (11). Inhibition of retinal axons by nonneuronal cells in the early chiasmatic pathway was advanced previously, but primarily as a partitioning device for the whole retinal projection, preventing retinal axons from growing into olfactory regions (58). Thus, axon guidance is likely to have multiple components: growth-supporting and growth-inhibiting, and substrate-bound and diffusible factors. How these cues operate in signaling growth cone navigation along paths to targets, particularly when changes in direction or divergence of subpopulations occurs, and to what extent these different factors are used in concert (63) are at present unclear.

Previous studies on retinal axon growth and guidance have suggested a number of mechanisms by which bilateral projections of ganglion cell axons are established. Spatial (10, 55), temporal (30, 66), activity-based (28), and even hormonal (31) factors have been proposed to play a role in divergence within the chiasm. Examination of the mutants in which retinal pigment epithelium is lacking (12, 23, 25) suggest that ganglion cell projection to one or the other side of the brain depends on properties of ganglion cell projection to one or the other side of the brain depends on properties of ganglion cells in the different regions. The two ganglion cell populations would thus be differentially labeled, allowing their axons to diverge. While the basis for the differential labeling has been probed and some dorsal–ventral (40, 65) or nasal–temporal differences (41) have been demonstrated, molecular identities of the crossed and uncrossed populations of retinal ganglion cells, usually arising from nasal and temporal retinal regions, respectively, have not as yet been established.

DYE-LABELING STUDIES IN FIXED BRAIN: AXON TRAJECTORY AND GROWTH CONE MORPHOLOGY REFLECT THE PATTERN OF DIVERGENCE

Labeling neurons in fixed brain with lipophilic dyes, in particular the carbocyanine dye DiI (1,1′-dioctadecyl-3,3,3′,3′-tetramethylindo-carbocyanine perchlorate) (18) provided an experimental approach to the issue of directed retinal axon growth or divergence, in embryonic mouse brain, at the time the retinal axons were extending toward their targets. To label the subpopulations of retinal neurons that give rise to crossed and uncrossed fibers, localized injections of the carbocyanine dye DiI were made in various parts of the retina, yielding labeling of a small number of fibers. Fibers labeled in the dorsal–nasal or temporal retina crossed the midline, whereas fibers labeled in the inferior temporal retina gave rise to uncrossed fibers (19).

The view of the trajectory of the two different populations of axons was striking. Ipsilateral-projecting axons from inferior temporal retina do not travel directly into the ipsilateral optic tract, nor do they travel separately or at different times than the contraterally projecting fibers (10, 53, 60). Instead, these fibers travel with crossing fibers toward the midline, then make a sharp turn back toward the ipsilateral side, generally within 150–200 μm proximal to the midline. Some fibers develop elaborate growth cones at the border of this region, appearing to extend along an unseen barrier (Fig. 1). Other fibers have growth cones which are in the act of turning away from the border of the midline region. The form of these elaborate growth cones closely resembles those seen *in vitro* along the border of a preferred and nonpreferred substrate (7, 36). These studies revealed that the divergence of crossed and uncrossed fibers occurs near the midline and implicated the midline as an inhibitory zone for uncrossed fibers, a zone containing cues that influence the laterality of projection of retinal axons.

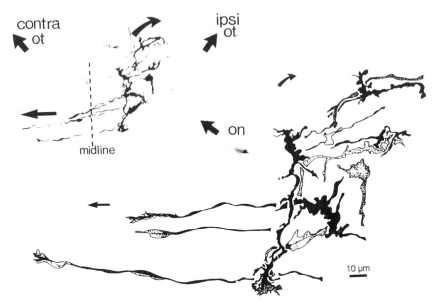

Fig. 1.   Growth cones of fibers with an uncrossed destination (from inferior temporal retina) become complex at the border of a zone along the midline. These fibers then turn back to the ipsilateral optic tract (ot, arrow in drawing). Some fibers from this part of the retina cross the midline. Note that the growth cones on the fibers at the midline zone border are more spread and irregular than those on the axons that grow straight across the midline. Crystals of DiI were placed in the retina of fixed mouse embryos at E16. Vibratome sections (75 μm) were then subjected to photoconversion, rendering the fluorescent label brown–black.

GROWTH CONE FORM AS AN INDICATOR OF BEHAVIOR AND CELL–CELL INTERACTIONS

Studies based on a number of different species have led to the principle that the form of growth cones changes as they grow through different locales along the path to targets (e.g., 3, 5, 8, 62, 68, and Fig. 1). These changes are thought to reflect responses to cellular and molecular cues. The results with DiI labeling in the fixed retina and brain highlight the finding that the most dramatic transitions, from more simple to expanded forms with numerous filopodia, occur at sites where growing neurites negotiate changes in direction. The particularly complex forms of growth cone at the border of this region suggested that they are perceiving cues important for the axon divergence. These observations are consistent with *in vitro* observations that growth cone shape transitions result from a change in behavior, either alterations of growth rates on varying substrates (1) or after interactions with different cell types (2, 32).

The combination of the pattern of axon projection and the dramatic changes in growth cone form then led us to examine three issues concerning cues for axon navigation in the chiasm: (i) the actual behaviors of axons in real time, to verify the behaviors deduced from fixed dye-labeled preparations; (ii) the cellular composition of the midline zone; and (iii) the growth cone interactions especially at the border of the midline zone.

REAL TIME STUDIES

To investigate the behaviors of retinal axons during the divergence, we developed a novel *in vitro* preparation in which the maneuvers of dye-labeled growth cones could be followed in the context of the intact living optic chiasm in real time, using low-light imaging (20, 21). After DiI is injected into selected ganglion cells in a semiintact preparation including the eye, optic nerves, and ventral slab of brain, the axons are viewed in the chiasm with fluorescence optics and an image-intensifier video camera. Time-lapse recordings are made over 3–20 hr, at both low and high power. Real-time video recordings of the behavior of living retinal ganglion cells have yielded remarkable views of how the divergence is accomplished and in general illuminate growth cone maneuvers in the context of the intact brain. This analysis also confirmed our studies on fixed brain that uncrossed fibers separate from crossed fibers near the chiasm midline. Fibers that remain ipsilateral attempt to penetrate the midline zone and develop complex growth cones which effect a sharp turn back to the ipsilateral side (Fig. 2).

Growth cones of fibers labeled in the inferior temporal retina extend up toward the midline and pause. At higher power, it is evident that they advance, then retract several times, cycling through spread shapes and branched forms, the latter extending along the boundary of the midline zone (20, 21). After several hours of this behavior, a "backward" filopodium forms, pointing toward the ipsilateral optic tract; a new growth cone develops at the tip of this filopodium and rapidly extends. The establishment of a new growth direction by a single filopodium has also been observed in the grasshopper nervous system (48). Thus, the video accounts indicated the sequence in which the different growth cone forms develop, in order to effect a turn, difficult to comprehend from static preparations, and provided additional sup-

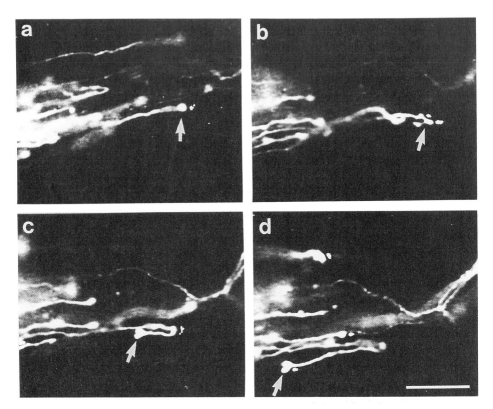

Fig. 2. The steps in turning of uncrossed fibers at the border of the midline zone, recorded over 8 hr with video microscopy and low-light imaging. Axons were labeled with DiI in the retina of an isolated preparation consisting of the retina, optic nerves, and a slab of the chiasm. In (a) the growth cone comes to a halt, then it spreads (b), and makes a turn (c). Bar: 50 μm.

port to the hypothesis that the midline zone is inhibitory to uncrossed fibers.

The video approach also revealed the kinetics of growth cone movement in the context of the intact nervous system. First, all fibers undergo periods of rapid growth, punctuated by long pauses. Second, these pauses often precede adjustments in the axis of growth, and appear to be more common near the midline, in fibers that cross as well as those that double back at the border of the midline zone. Third, shifts in the axis of growth are heralded by the extension of lamellipodia or filopodia, which then define the new growth axis as organelles fill the newly formed extension. Finally, distinct forms are common during rapid advance and pauses, the more streamlined forms prominent during advance, and complex forms during pauses. These observations agree with other recent characterizations of growth cone behavior in other intact or semiintact vertebrate nervous systems (29, 49, 61).

MIDLINE OF THE OPTIC CHIASM: A FIRST CUE FOR DIVERGENCE

To understand what cues might be present at the midline, the point in the path where the most dramatic shape transformations occur as fibers turn back, we investigated the cellular composition of this locale by immunohistochemistry and the monoclonal antibody RC2, a ligand selective for immature astroglial cells in the mouse central nervous system (CNS) (43). These studies revealed a palisade of radial glial cells that curves from the bottom of the third ventricle around the midline, spanning 100–200 μm to either side of the midline and corresponding to the zone near the midline which uncrossed fibers never penetrate (37, 39, and Fig. 3). While some RC2-positive cells are seen in the lateral chiasm, these cells have stellate shapes and are sparse. At the posterior chiasm border and anterior to the chiasm, radial glia fan out from the ventricle as an unbroken curtain. Thus, in the anterior portion of the chiasm, where the newly arrived fibers project across the midline, the radial glial formation is restricted to the midline sector.

In sections double-labeled for RC2 and DiI, crossing axons pass through the RC2-positive palisade and ipsilaterally projecting axons make their turn at the border of the palisade (37, 39). Ultrastructural analysis of preparations in which growth cones were labeled with DiI and then followed through thin sections show that complex growth cones or turning growth cones abut the radial glial fibers or send their projections among them, but the entire growth cone is never within the glial palisade. In contrast, growth cones that are seen crossing the midline, with simpler shapes, are interposed in their entirety along groupings of radial fibers. Radial glial endfeet are prominent in the nerve-chiasm path (26, 57) and glial filament protein-positive cells described in the chiasm from neonatal periods onward (4), but the radial glial midline palisade has not been previously observed. The immunocytochemical analysis demonstrated that the optic chiasm contains a midline structure composed of specialized glia that appears to

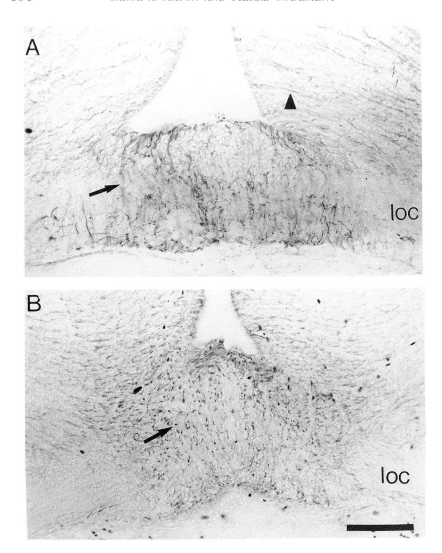

Fig. 3. Radial glial fibers fill a zone adjacent to both sides of the midline of the optic chiasm. (A) Frontal cryostat section, immunostained with the peroxidase method and a monoclonal to mouse radial glia, RC2. Radial glial fibers extend from the floor of the third ventricle to the pia along the floor of the chiasm (arrow). The lateral chiasm (loc) lacks such fibers, but radial glia elsewhere in the brain are also stained (arrowhead). (B) Horizontal section, prepared in the same manner as in (A), showing radial fibers cut in cross section, appearing as dots. Bar: 10 μm.

play a dual role in axon guidance, both permissive and inhibitory (Fig. 4, step 1).

Specialized glial cells at the midline have been implicated in axon guidance in a wide variety of species and points in the neuraxis. Glial cells in the midline of insect CNS play a role in crossing of certain axons (33). The floor plate of the chick and mouse spinal cord acts as a permissive crossing thoroughfare for commissural axons (5, 15, 68). The floor plate in zebrafish spinal cord, like the chiasm midline, plays a dual role, acting as a cross point for commissural fibers, but a barrier for other populations (34). In several cases, midline structures such as the dorsal roof plate (59) may be exclusively nonpermissive for axon growth across the midline. Further support for these specialized midline sectors acting as guidance cues comes from experiments in which disruption of identified glial cells in the midline of the CNS results in aberrant axonal projections (6, 9, 56).

FIBER–FIBER INTERACTIONS ALSO CONTRIBUTE TO AXON DIVERGENCE

The midline glial structure is one possible cue for the turning of uncrossed fibers (Fig. 4, step 1). A second is fiber–fiber interactions (Fig. 4, step 2). Cues could reside on the surfaces of other axons extending through the pathway (52). Interactions between nasal and temporal fibers themselves have been invoked previously as a primary

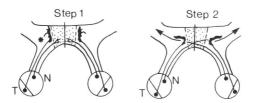

Fig. 4. A two-step mechanism for axon divergence in the optic chiasm. Step 1: while fibers from dorsal nasal (N) or temporal retina cross the chiasm midline, fibers from the inferior temporal retina (T) approach the midline, but do not cross the midline zone (dashed line, dots represent radial glia). Growth cones spread along this border, and a "backward" filopodium is extended (asterisk). Step 2: The backward filopodium contacts a fiber from the other eye and fasciculates with it, fully effecting the turn away from the midline zone.

mechanism in retinal axon divergence in rodents (17). Such interactions would include fiber avoidance as well as fasciculation.

By investigating generation of crossed and uncrossed projections in the mouse visual pathways with retrograde tracing methods, the characteristics and numbers of retinal ganglion cells within the uncrossed projections were determined in mice in which one eye had been removed early in embryonic development (17). These studies showed that the retinal origin of these projections is grossly normal, although the numbers of cells projecting ipsilaterally is reduced. This pattern is similar to that in mice (and other species) with reduced pigment in the retinal epithelium (12, 23, 25).

Labeling different retinal regions in fixed brain with DiI at E16 after enucleation at E13 implicate fiber–fiber interactions. Fibers from inferior temporal retina that would normally turn at the midline simply approach it and stall here (19). Because these fibers do not aberrantly cross, or grow back normally in the absence of crossing fibers from the other eye, a two-step mechanism for guidance of uncrossed fibers is suggested, a cue at the midline impeding growth of uncrossed fibers and a cue for growth with fibers from the other eye once the turn is made. This strengthened the hypothesis that fiber–fiber interactions, between fibers from each eye, could be involved in the events during crossing at the optic chiasm, perhaps by selective fasciculation of like fibers.

The analysis of retinal axon divergence in the optic chiasm with DiI tracing in living brain in preparations lacking one eye further strengthens this hypothesis (20, 21). Fibers from the inferior temporal retina approach the midline and accumulate at the border of the midline zone, neither turning nor crossing abnormally. These real-time observations of uncrossed axon behavior in enucleated animals demonstrate that subsequent to responding to the inhibitory cue intrinsic to the midline glial structure (Fig. 4, step 1), ipsilateral-projecting fibers depend on interactions with crossing fibers from the opposite eye (Fig. 4, step 2). Such a two-step mechanism, interaction with nonneuronal cues providing directionality information for crossing and neuron–neuron association to complete crossing, has also been proposed for insect neurons in CNS ganglia (45).

In summary, we have identified the trajectory and growth cone interactions of retinal axons, focusing on the midline of the optic chiasm, where fibers diverge to each side of the brain. Fibers from dorsal temporal and nasal retina, the source of crossed fibers, traverse the midline of the optic chiasm. In contrast, fibers originating in inferior temporal retina, the source of the permanent uncrossed projection growing through the chiasm at E15–17, approach the midline, develop highly complex and branched growth cones, and make a sharp turn back to the ipsilateral tract.

The change of course for inferior temporal fibers is abrupt at the border of the midline zone, which is occupied by a palisade of radial glia. The cells of the midline, principally the radial glia, might express molecules to which nasal and temporal growth cones respond differentially, molecules which could prevent ipsilateral axons from crossing the midline region. Crossing fibers also appear to pause near the midline, indicating that the midline zone represents a transition to a different and perhaps difficult zone for both sets of fibers to traverse. Moreover, after unilateral eye removal, fibers that would otherwise turn, stall at the midline zone, indicating that subsequent to inhibition by the midline, uncrossed fibers interact with fibers from the other eye.

CONCLUDING REMARKS

Our analysis of retinal projections through the chiasm points to a role for the midline glial cells in patterning the bilateral organization of retinal pathways. To further test for permissive and inhibitory cellular components of the optic chiasm midline, we have cocultured dissociated cells of the optic chiasm with retinal explants (22). These cells contain the midline radial glia, precursor cells from the ventricultar zone of the third ventricle, and other nonneuronal cells. Fibers from explants of inferior temporal retina, which give rise to uncrossed axons, turn away from the border of fields of cells from the chiasm midline, whereas axons from dorsal temporal or dorsal nasal retina grow freely on these fields. When plated directly on dissociated chiasm cells (22) or on membranes of chiasm cells (67), fibers from explants of inferior temporal retina grow shorter neurites than those extending from explants of dorsal temporal retina, a source of crossed fibers. These results confirm that the midline region plays a dual role in growth support of retinal axons but leaves open the question as to which is the critical cell population inhibiting the uncrossed fibers. In addition, two remaining questions are the molecules expressed by the midline relevant to the divergence and the molecular basis of the differential response of the two retinal ganglion cell populations.

Because of the emerging theme of midline structures in mediating the bilateral organization of crossed pathways in a number of species, it will now be of interest to compare the chiasm midline, both its molecular properties and its origins, with those in other species or in other parts of the mammalian CNS. The analogies in location and guidance role shared by the chiasm midline and spinal cord floor plate, i.e., they are both midline structures that serve as passage ways for crossing axons, led us to investigate whether the chiasm expresses epitopes similar to those expressed by floor plate cells. Among them is a marker for stage-specific embryonic antigen expressed on embryonic stem cells (SSEA I) (13, 37). Antibodies to this molecule stain some but not all fibers of the glial palisade. Moreover, while monoclonal antisera to RC2 reveals the chiasm midline glial structure and the floor plate, it does not differentiate between these glia and other radial glia. In contrast, the markers for SSEA I and other markers that reveal the floor plate also selectively stain the chiasm midline.

Another emerging issue is how crossed pathways such as the optic chiasm are established early in development. Little is known about the specification of this region, or the timing of its establishment relative to retinal axon outgrowth. Kimmel and co-workers (64) have raised the possibility that crossed pathways in the hindbrain of zebrafish may occur at segment boundaries, bounded on either side by glial septa or "curtains" stretching from side to side. The chiasm occurs at the merger of the diencephalon and telencephalon, and the chiasm midline palisade is sandwiched in between two such glial curtains. While segmentation per se is not obvious in the forebrain, it is becoming apparent that a number of genes with analogies to those in lower animals regulating segmentation and occurring at segment boundaries are expressed in mouse brain [see Fritsch and Gruss (Chapter 16) and Hunt and Krumlauf (Chapter 15)], some expressed with the optic chiasm (51). With these probes, it should be possible to address the mechanisms regulating partitioning of the neuraxis, in the anteroposterior axis as well at the midline, that ultimately specify axon projection patterns.

ACKNOWLEDGMENTS

We are grateful to our colleagues Drs. Li-Chong Wang, Gord Fishell, Mary Beth Hatten, and Martin Seidensticker for many insightful discussions. We thank Kenneth Wei and Sue Rosenstock for assistance in word processing and Raynard Manson for photographic assistance. Supported by NIH Grant NS 27615 and NATO Grant 890370.

REFERENCES

1. Argiro, V., Bunge, M. B., and Johnson, M. I. (1984). Correlation between growth cone form and movement and their dependence on neuronal age. *J. Neurosci.* **4**, 3051–3062.

2. Bandtlow, C., Zachleder, T., and Schwab, M. E. (1990). Oligodendrocytes arrest neurite growth by contact inhibition. *J. Neurosci.* **10**, 3837–3848.

3. Bovolenta, P., and Mason, C. A. (1987). Growth cone morphology varies with position in the developing mouse visual pathway from retina to first targets. *J. Neurosci.* **7**, 1447–1460.

4. Bovolenta, P., Liem, R. K. H., and Mason, C. A. (1987). Onset of glial filament protein expression and development of astroglial shape in the mouse retinal axon pathway. *Dev. Brain Res.* **33**, 113–126.

5. Bovolenta, P., and Dodd, J. (1990). Guidance of commissural growth cones at the floor plate in embryonic rat spinal cord. *Development* **109**, 435–438.

6. Bovolenta, P., and Dodd, J. (1991). Perturbation of neuronal differentiation and axon guidance in the spinal cord of mouse embryos lacking a floor plate: Analysis of Danforth's short-tail mutation. *Development* **113**, 625–639.

7. Burmeister, D. W., and Goldberg, D. J. (1988). Micropruning: The mechanism of turning of *Aplysia* growth cones at substrate borders *in vitro*. *J. Neurosci.* **8**, 3151–3159.

8. Caudy, M., and Bentley, D. (1986). Pioneer growth cone morphologies reveal proximal increases in substrate affinity. *J. Neurosci.* **6**, 364–379.

9. Chitnis, A. B., and Kuwada, J. Y. (1991). Elimination of brain tract increases errors in pathfinding by follower growth cones in the zebrafish embryo. *Neuron* **7**, 277–285.

10. Colello, R. J., and Guillery, R. W. (1990). The early development of retinal ganglion cells with uncrossed axons in the mouse: Retinal position and axon course. *Development* **108**, 515–523.

11. Cox, E. C., Muller, B., and Bonhoeffer, F. (1990). Axonal guidance in the chick visual system: Posterior tectal membranes induce collapse of growth cones from temporal retina. *Neuron* **4**, 31–37.

12. Cucchiaro, J. B. (1991). Early development of the retinal line of decussation in normal and albino ferrets. *J. Comp. Neurol.* **312**, 193–207.

13. Dodd, J., and Jessell, T. M. (1985). Lactoseries carbohydrates specify subsets of dorsal root ganglion neurons projecting to the superficial dorsal horn of rat spinal cord. *J. Neurosci.* **5**, 3278–3294.

14. Dodd, J., and Jessell, T. J. (1988). Axonal guidance and the patterning of neuronal projections in vertebrates. *Science* **242**, 692–699.

15. Dodd, J., Morton, S. B., Karagogeos, D., Yamamoto, M., and Jessell, T. M. (1988). Spatial regulation of axonal glycoprotein expression on subsets of embryonic spinal neurons. *Neuron* **1**, 105–116.

16. Dräger, U. C. (1985). Birth dates of retinal ganglion cells giving rise to the crossed and uncrossed optic projections in the mouse. *Proc. R. Soc. Lond. B* **224**, 57–77.

17. Godement, P., Salaün, J., and Métin, C. (1987). Fate of uncrossed retinal projections following early or late prenatal monocular enucleation in the mouse. *J. Comp. Neurol.* **255**, 97–109.

18. Godement, P., Vanselow, J., Thanos, S., and Bonhoeffer, F. (1987). A study in

developing visual systems with a new method of staining neurones and their processes in fixed tissue. *Development* **101**, 697–713.

19. Godement, P., Salaün, J., and Mason, C. A. (1990). Retinal axon pathfinding in the optic chiasm: Divergence of crossed and uncrossed fibers. *Neuron* **5**, 173–196.

20. Godement, P., and Mason, C. A. (1990). Behavior of live retinal axon growth cones in the optic chiasm. *Soc. Neurosci. Abstr.* **16**, 1125.

21. Godement, P., and Mason, C. A. Retinal axon navigation in the optic chiasm: growth cone behavior and midline cues, submitted for publication.

22. Guillaume, R., Wang, L.-C., Mason, C. A., and Godement, P. (1991). Retinal axon-optic chiasm cellular interactions *in vitro*. *Soc. Neurosci. Abstr.* **17**, 40.

23. Guillery, R. W. (1974). Visual pathways of albinos. *Sci. Am.* **230**, 44–54.

24. Guillery, R. W. (1982). The optic chiasm of the vertebrate brain. *Contrib. Sensory Physiol.* **7**, 39–73.

25. Guillery, R. W. (1989). Early monocular enucleations in fetal ferrets produce a decrease of uncrossed and an increase of crossed retinofugal components: A possible model for the albino abnormality. *J. Anat.* **164**, 73–84.

26. Guillery, R. W., and Walsh, C. (1987). Changing glial organization relates to changing fiber order in the developing optic nerve of ferrets. *J. Comp. Neurol.* **265**, 203–217.

27. Harrelson, A. L., and Goodman, C. S. (1988). Growth cone guidance in insects: Fasciclin II is a member of the immunoglobulin superfamily. *Science* **242**, 700–708.

28. Harris, W. A. (1984). Axonal pathfinding in the absence of pathways and impulse activity. *J. Neurosci.* **4**, 1153–1162.

29. Harris, W. A., Holt, C. E., and Bonhoeffer, F. (1987). Retinal axons with and without their somata, growing to and arborizing in the tectum of frog embryos: A time-lapse video study of single fibers *in vivo*. *Development* **101**, 23–133.

30. Holt, C. E. (1984). Does timing of axon outgrowth influence initial retinotectal topgraphy in *Xenopus?* *J. Neurosci.* **4**, 1130–1152.

31. Hoskins, S. G., and Grobstein, P. (1985). Development of the ipsilateral retinothalamic projection in the frog *Xenopus laevis:* Role of thyroxine. *J. Neurosci.* **5**, 930–940.

32. Kapfhammer, J., and Raper, J. (1987). Collapse of growth cone structure on contact with specific neurites in culture. *J. Neurosci.* **7**, 201–212.

33. Klambt, C., Jacobs, J. R., and Goodman, C. S. (1991). The midline of the *Drosophila* central nervous system: A model for the genetic analysis of cell fate, cell migration, and growth cone guidance. *Cell* **64**, 801–815.

34. Kuwada, J. Y., Bernhardt, R. R., and Chitnis, A. B. (1990). Pathfinding by identified growth cones in the spinal cord of zebrafish embryos. *J. Neurosci.* **10**, 1299–1308.

35. Landmesser, L. (1991). Growth cone guidance in the avian limb: A search for cellular and molecular mechanisms. *In* "The Nerve Growth Cone" (S. B. Kater, P. Letourneau, and E. Macagno, eds.), pp. 373–385. Raven Press, New York.

36. Letourneau, P. (1975). Cell-to-substratum adhesion and guidance of axonal elongation. *Dev. Biol.* **44**, 92–101.

37. Mason, C. A., Blazeski, R., Misson, J.-P., and Godement, P. (1990). *Soc. Neurosci. Abstr.* **16**, 1125.

38. Mason, C. A., Dodd, J., Blazeski, R., and Godement, P. (1991). The midline of the mouse optic chiasm: Cellular composition and antigen expression during retinal axon growth. *Soc. Neurosci. Abstr.* **17**, 39.

39. Mason, C. A., Blazeski, R., and Godement, P. The midline of the optic chiasm: Growth cone-radial glial interactions, submitted for publication.

40. McCaffery, P., Tempst, P., Lara, G., and Drager, U. C. (1991). Aldehyde dehydrogenase is a positional marker in the retina. *Development* **112**, 693–702.

41. McLoon, S. (1991). A monoclonal antibody that distinguishes between temporal and nasal retinal axons. *J. Neurosci.* **11**, 1470–1477.

42. Métin, C., Godement, P., and Imbert, M. (1988). The primary visual cortex of the mouse: Receptive field properties and functional organization. *Exp. Brain Res.* **69**, 594–612.

43. Misson, J.-P., Edwards, M. A., Yamamoto, M., and Caviness, V. S. (1988). Identification of radial glial cells within the developing murine cerebral wall: Studies based upon a new histochemical marker. *Dev. Brain Res.* **44**, 95–108.

44. Mori, K., Ikeda, J., and Hayaishi, O. (1990). Monoclonal antibody R2D5 reveals midsagittal radial glial system in postnatally developing and adult brainstem. *Proc. Natl. Acad. Sci. U.S.A.* **87**, 5489–5493.

45. Myers, P. Z., and Bastiani, M. J. (1991). *In vivo* time-lapse video microscopy of commissural growth cones in the grasshopper CNS. *Soc. Neurosci. Abstr.* **17**, 533.

46. Nornes, H. O., Dressler, G. R., Knapik, E. W., Deutsch, U., and Gruss, P. (1990). Spatially and temporally restricted expression of Pax2 during murine neurogenesis. *Development* **109**, 797–809.

47. Oakley, R. A., and Tosney, K. W. (1991). Peanut agglutinin and chondroitin-6-sulfate are molecular markers for tissues that act as barriers to axon advance in the avian embryo. *Dev. Biol.* **147**, 187–206.

48. O'Connor, T. P., Duerr, J. S., and Bentley, D. (1990). Pioneer growth cone steering decisions mediated by single filopodial contacts in situ. *J. Neurosci.* **10**, 3935–3946.

49. O'Rourke, N. A., and Fraser, S. E. (1990). Dynamic changes in optic fiber terminal arbors lead to retinotopic map formation: An in vivo confocal microscopic study. *Neuron* **5**, 159–171.

50. Patterson, P. H. (1988). On the importance of being inhibited, or saying no to growth cones. *Neuron* **1**, 263–267.

51. Price, M., Lemaistre, M., Pischetola, M., Di Lauro, R., and Duboule, D. (1991). A mouse gene related to Distal-less shows a restricted expression in the developing forebrain. *Nature* **351**, 748–751.

52. Raper, J. A., and Grunewald, E. B. (1990). Temporal retinal growth cones collapse on contact with nasal retinal axons. *Exp. Neurol.* **109**, 70–74.

53. Reese, B. E., Guillery, R. W., Marzi, C. A., and Tassinari, G. (1991). Position of axons in the cat's optic tract in relation to their retinal origin and chiasmatic pathway. *J. Comp. Neurol.* **306**, 539–553.

54. Reichardt, L. F., Bossy, B., Carbonetto, S., DeCurtis, I., Emmett, Hall, D. E., and Ignatius, M. J. (1991). Neuronal receptors that regulate axon growth. *Cold Spring Harbor Symp. Quant. Biol.* **55**, 341–350.

55. Silver, J., and Sidman, R. C. (1980). A mechanism for the guidance and topographic patterning of retinal ganglion cell axons. *J. Comp. Neurol.* **189**, 101–111.

56. Seeger, M. A., Tear, G., Ferres-Marco, D., and Goodman, C. S. (1991). *Commissureless*, a mutation in Drosophila that specifically disrupts growth cone guidance towards the midline. *Soc. Neurosci. Abstr.* **17**, 742.

57. Silver, J., and Rutishauser, U. (1984). Guidance of optic axons *in vivo* by a preformed adhesive pathway on neuroepithelial endfeet. *Dev. Biol.* **106**, 485–499.

58. Silver, J., Poston, M., and Rutishauser, U. (1987). Axon pathway boundaries in the

developing brain. I. Cellular and molecular determinants that separate the optic and olfactory projections. *J. Neurosci.* **7**, 2264–2272.

59. Snow, D. M., Steindler, D. A., and Silver, J. (1991). Molecular and cellular characterization of the glial roof plate of the spinal cord and optic tectum: A possible role for a proteoglycan in the development of an axon barrier. *Dev. Biol.* **138**, 359–376.

60. Sretavan, D. W. (1990). Specific routing of retinal ganglion cell axons at the mammalian optic chiasm during embryonic development. *J. Neurosci.* **10**, 1995–2007.

61. Sretavan, D. W. (1990). Axon navigation at the mammalian optic chiasm: direct observation using fluorescent time-lapse video microscopy. *Soc. Neurosci. Abstr.* **16**, 1125.

62. Tosney, K. W., and Landmesser, L. T. (1985). Growth cone morphology and trajectory in the lumbosacral region of the chick embryo. *J. Neurosci.* **5**, 2345–2358.

63. Tosney, K. W. (1991). Growth cone navigation in the proximal environment of the chick embryo. *In* "The Nerve Growth Cone" (S. Kater, P. Letourneau, and E. Macagno, eds.), pp. 387–403. Raven Press, New York.

64. Trevarrow, B., Marks, D. L., and Kimmel, C. B. (1990). Organization of hindbrain segments in the zebrafish embryo. *Neuron* **4**, 669–679.

65. Trisler, G., Schneider, M. D., and Nirenberg, M. (1981). A topographic gradient of molecules in retina can be used to identify neuron position. *Proc. Natl. Acad. Sci. U.S.A.* **78**, 2145.

66. Walsh, C. (1986). Age-related fiber order in the ferret's optic nerve and optic chiasm. *J. Neurosci.* **6**, 1635–1642.

67. Wizenmann, A., Thanos, S., and Bonhoeffer, F. (1991). Behavior of retinal fibres *in vitro* confronted with membranes from the chiasma midline. *Eur. J. Neurosci.* **4** (Suppl.), 93.

68. Yaginuma, H., Shunsaku, H., Kunzi, R., and Oppenheim, R. W. (1991). Pathfinding by growth cones of commissural interneurons in the chick embryo spinal cord; a light and electron microscopic study. *J. Comp. Neurol.* **304**, 78–102.

8

Cellular Interactions Regulating the Formation of Terminal Arbors by Primary Motoneurons in the Zebrafish

MONTE WESTERFIELD AND DENNIS W. LIU
Institute of Neuroscience
University of Oregon
Eugene, Oregon 97403

To understand how the nervous system develops, we need to know how specific synaptic connections form between neurons and their targets. To make correct connections, neurons must extend neurites to an appropriate region and then form synapses with the correct type and number of target cells. The mechanisms that regulate the various steps involved in this process are still virtually unknown.

One approach which has been useful in recent years for studying synapse formation is to study single "identified" neurons that can be recognized in different individuals of a given species. By examining the growth of these neurons in various experimental situations, it is possible to understand something about the factors that regulate their development. In general, this type of analysis has been conducted primarily on invertebrate species whose nervous systems are composed of relatively few neurons, many of which can be uniquely identified. However, recent studies of one vertebrate, the embryonic zebrafish, have shown that a number of early developing "primary neurons" can be uniquely identified and studied throughout early development (1).

Cell–Cell Signaling in
Vertebrate Development

In both types of animals, considerable progress has been made toward understanding the early steps in the formation of specific synaptic connections. The earliest neurons to differentiate, often termed pioneers, extend growth cones along pathways which later become the tracts of the central nervous system or the nerves of the periphery. In insects, initial pathfinding by these pioneer neurons is highly stereotyped (2–6) and in cases where the final specificity of synaptic connections is known, growth cones extend directly toward appropriate targets (5, 6). Evidence from chicks (7) and frogs (8, 9) suggests that the growth cones of spinal motoneurons grow directly to the regions where their appropriate targets reside and, if displaced experimentally, they will take alternative routes to reach appropriate targets (7). In the zebrafish, we have observed precise cell-specific pathfinding by the growth cones of individually identified motoneurons in live embryos (10, 11). These growth cones follow highly stereotyped, cell-specific pathways to reach their targets.

IDENTIFIED PRIMARY MOTONEURONS OF THE ZEBRAFISH

The fast twitch, white muscle fibers on each side of each body segment of the zebrafish are innervated by three or four uniquely identifiable primary motoneurons (12, 13). These motoneurons are characteristically located in discrete positions along the rostral–caudal axis of the spinal segment. The caudalmost cell (named CaP, for caudal primary) innervates muscle fibers exclusively in the ventral region of the segment. The middle cell (named MiP, for middle primary) innervates fibers in the dorsal region of the segment and the rostralmost cell (named RoP, for rostral primary) innervates fibers in the central region of the segment between the territories of the other two primary motoneurons. The fourth motoneuron (named VaP, for variable primary) is located next to the CaP motoneuron but is present in only some segments (13). It takes on a variable fate; in some cases it innervates an intermediate region of the myotome and in others it dies. In the adult, each muscle fiber is innervated by a single primary motoneuron and within a given muscle region the corresponding primary motoneuron innervates all the fibers (12). There are also a number of smaller, secondary motoneurons that develop later and share innervation of these muscle fibers with the primary motoneurons.

PATHFINDING BY THE PRIMARY MOTONEURONS

The primary motoneurons follow cell-specific pathways. CaP is always the first motoneuron in each segment to sprout a growth cone and to extend a process out of the spinal cord (10, 11). It is followed by MiP and then by RoP. All three growth cones initially extend to the region of the horizontal myoseptum, a structure that separates the dorsal and ventral muscles of the segment. The growth cones then diverge and take cell-specific pathways toward the regions of muscle they will ultimately innervate. Recent studies demonstrated that this directed outgrowth, or pathfinding, occurs normally in the absence of potential competitive interactions from neighboring primary motoneurons (14, 15). Cell-specific pathfinding has been offered as a partial explanation of the specificity of innervation since, from the outset, the growth cone of each motoneuron characteristically extends into an appropriate region of muscle rather than extending tentative, inappropriate projections which must be retracted during subsequent development (16).

FORMATION OF TERMINAL ARBORS

Correct pathfinding can lead a growth cone to the proper target region, but pathfinding alone cannot account for the precise shape and size of the terminal field of a neuron. Once a growth cone reaches the appropriate target region, there must be mechanisms which constrain its growth and determine its choice of synaptic partners. Growth cones may be prevented from extending into inappropriate regions, as seems to be the case for the segmental specificity of zebrafish primary motoneurons which do not cross the borders between muscle segments (11, 17, 18). On the other hand, motor axons in most vertebrates initially branch exuberantly in target regions and later eliminate redundant projections (19). In embryonic mammals and birds, and in larval amphibians, many motoneurons die, including some whose axons have reached the target region (20, 21). During later developmental stages large numbers of neuromuscular synapses are eliminated (22). Regressive events also occur in invertebrate nervous systems in which they may play a role in establishing synaptic specificity (23, 24). In most systems, regressive events seem to be regulated by activity-dependent competitive interactions among neurons (25–27).

We examined the formation of terminal arbors by individual primary motoneurons in zebrafish embryos and larvae to study the relative roles of directed growth and synapse elimination in the establishment of terminal arbors and to assess the influence of activity-dependent competitive interactions in regulating the shape and size of axonal arbors. To do this we labeled individual motoneurons with long-lasting fluorescent dyes and directly observed their growth during the first week of development when their terminal arbors form (28).

We found that, as seen in previous studies (16), the growth cone of each primary motoneuron migrated out of the spinal cord and after reaching the horizontal myoseptum extended directly toward its appropriate target region. In the case of the CaP and MiP axons which innervate ventral and dorsal muscles, respectively, we consistently observed that the growth cones bifurcated when they reached the ventral or dorsal edge of the somite (Fig. 1). Subsequent to bifurcation, one of the resulting growth cones formed a branch which grew caudally into the muscle while the other branch extended laterally and into the region of the rostral segment border. All other second-order branches grew directly from the main axonal trunk usually by sprouting from varicosities left behind by the growth cone. This pattern of growth was repeated during the first few days of development, with higher-order branches forming primarily from varicosities left behind the growth cones rather than from bifurcation of the growth cones themselves. Although neuronal branches in cell culture form most frequently by growth cone bifurcation, terminal arbors *in vivo* may develop more commonly by this form of sprouting which has also been observed in the *Xenopus* (29) and mammalian (30) central nervous systems.

EXTENSION AND RETRACTION OF PROJECTIONS

The CaP motoneuron often extended one or a few branches into the muscle near the horizontal myoseptum, a region which is ultimately innervated by the RoP motoneuron. These septal branches probably formed functional, neuromuscular junctions with muscle fibers in this region because acetylcholine receptors clustered on the surfaces of the muscle fibers underlying these branches (31, 32) and because cholinergic synaptic transmission occurred (33, 31). In every case, these CaP branches retracted around the time that the RoP motoneuron extended branches into this region. To see if there was actually a period of time when the two arbors overlapped, we labeled both the

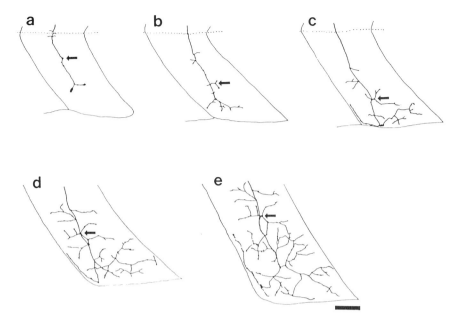

Fig. 1. Side branches sprout from existing varicosities and continue to grow throughout the observation period. (a–e) Taken from sequential observations of a single labeled CaP motoneuron at (a) 24 hr, (b) 30 hr, (c) 43 hr, (d) 57 hr, and (e) 69 hr after fertilization. The axonal arbor is drawn in heavy solid lines, the dotted line represents the horizontal myoseptum, and the outer lines indicate the segment boundaries. The arrows point to a varicosity, from which side branches grew, as it appeared at each time point. Sagittal view of segment 8 on the right-hand side, dorsal to the top and rostral to the left. Bar: 15 μm. (Adapted from Liu and Westerfield, 28.)

CaP and RoP motoneurons in the same segment with different colored fluorescent dyes. We found that individual branches from the two motoneurons, in some cases, extended along the same muscle fiber although they did not always occupy precisely the same location on the fiber (Fig. 2). By labeling acetylcholine receptors with a fluorescent conjugate of α-bungarotoxin (28, 32), we saw that separate clusters of receptors formed under individual branches from each of the two primary motoneurons, consistent with the notion that both motoneurons innervate the same fiber. During the following few hours, the CaP branches consistently retracted from fibers within this region, while the RoP branches continued to grow (Fig. 3). In most cases, acetylcholine receptor clusters remained on the surface of the muscle fiber for a few hours after retraction of the CaP branches before they, too,

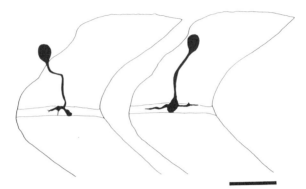

Fig. 2.   Early CaP and RoP branches contact the same muscle fiber. Drawing of the RoP (left) and CaP (right) motoneurons labeled in the same segment as they appeared at 21 hr after fertilization. The RoP motoneuron had extended a branch rostrally along the muscle cell, outline shown by the horizontal lines. At the same time, the CaP motoneuron in this segment had extended long branches both rostrally and caudally along the same muscle fiber. Sagittal view of segment 12 on the right-hand side, dorsal to the top, rostral to the left. Bar: 20 μm. (Adapted from Liu and Westerfield, 28.)

were eliminated. This observation suggests that competition for receptors is probably not a regulator of synapse elimination in this system as has been suggested from studies of mammals (34).

We wondered if the retraction of the misplaced CaP branches might be triggered by arrival of appropriate branches from the RoP motoneuron. To test this idea, RoP (and MiP) motoneurons were ablated by laser irradiation (14) before they extended growth cones out of the spinal cord and the subsequent development of the surviving CaP motoneuron was followed (28). In every case in which CaP extended branches into the region of the horizontal myoseptum, these branches were later retracted in a normal manner even though the other primary motoneurons were absent. Thus, it seems unlikely that competitive interactions with RoP or MiP regulate this aspect of the synaptic specificity of the CaP motoneuron.

Another way that the retraction of CaP branches might be regulated is by activity-dependent interactions with the muscles or later developing motoneurons. We examined these possibilities by pharmacologically blocking neuromuscular transmission and by studying mutant zebrafish, *nic-1* (b107) (32), which lack functional acetylcholine receptors (28). All three blockers, tricaine, curare, and α-bungarotoxin, had no effect on outgrowth, terminal formation, or retraction of branches. Similarly, branches grew and retracted normally in *nic-1*

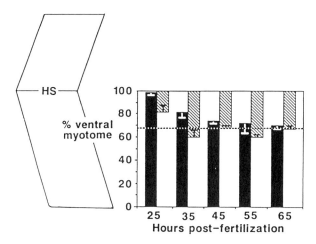

Fig. 3. CaP and RoP axonal arbors transiently overlap. The locations of axonal branches of individually labeled CaP and RoP motoneurons were observed at different developmental times from the onset of axogenesis (about 17 hr after fertilization) up to 2.5 days of development. The average locations of the most dorsal CaP axonal branch (solid bars) and the most ventral RoP branch (hatched bars) were calculated as a percentage of the distance between the horizontal myoseptum (100%) and the ventral border of the muscle segment (0%) and were plotted as functions of developmental age, "hours postfertilization." Initially, the most dorsal CaP branches were located at the horizontal myoseptum (99% at 25 hr). During subsequent development these branches retracted so that, by approximately 65 hr, the most dorsal branch was found about a third of the way toward the ventral edge of the muscle (70%) similar to the location occupied by adult CaP branches (broken line). During the same time period, RoP branches extended farther ventrally from the region of the horizontal myoseptum. During the first 2 days, the most dorsal CaP branch was closer to the horizontal myoseptum than the most ventral RoP branch and thus produced significant overlap between the CaP and RoP arbors. A schematic drawing of a muscle segment is shown on the left for orientation. HS, horizontal myoseptum. Data are from 98 observations of 41 CaP motoneurons in 41 animals and 39 observations of 11 RoP motoneurons in 11 animals. (From Liu and Westerfield, 28.)

mutant embryos. Moreover, the shape and size of the CaP terminal arbor was normal in these animals. Thus, this system appears to develop normally in the absence of activity-dependent interactions.

These observations demonstrate that, like other vertebrate motoneurons, zebrafish primary motoneurons consistently produce and later eliminate extra synapses. However, they also suggest that there is a fundamental difference between primary motoneurons and later developing motoneurons in terms of the mechanisms that regulate the elimination of these redundant synapses. Perhaps primary motoneurons depend upon specific substrate or cellular interactions (17, 18) to pioneer the peripheral nerves and do not need activity-dependent mechanisms to regulate their synaptic specificity.

REGULATION OF BRANCH EXTENSION AND RETRACTION

What, then, is the function of the extension and retraction of septal branches? The highly stereotyped sequence of these events may indicate that important interactions are occurring between the growth cones and something near the horizontal myoseptum. It may be that this region is particularly attractive for growth cones and that these same properties induce and transiently stabilize branches. Similar growth cone behavior has been observed in insects when pioneer sensory growth cones contact guidepost cells (35) or when pioneer motor growth cones encounter segment boundaries (36) or muscle pioneers (5). According to this view, the placement of CaP branches in the region of the horizontal myoseptum is not strictly an error in neuronal projection, but rather reflects an important step in axonal outgrowth and expression of cellular identity.

SUMMARY

Studies of the embryonic zebrafish have allowed characterization of the development of individually identified primary motoneurons. The growth cones of these neurons pioneer the pathways that later are followed by other axons that comprise the major peripheral nerves. Each primary motoneuron exclusively innervates a specific subset of axial muscle fibers within the corresponding body segment. The specificity of this innervation is, in part, established by cell-specific pathfinding by the growth cones of these neurons. Each growth cone follows a particular pathway and extends directly toward the region of muscle containing appropriate targets.

The precise numbers and positions of the muscle fibers innervated by the primary motoneurons are determined by the shape and size of the neuronal arbors. The arbors develop primarily by the sprouting of second-order branches from axonal varicosities, rather than by bifurcation of the growth cones. With a few notable exceptions, nearly all second and higher-order branches extend exclusively into appropriate muscle regions. The CaP motoneuron, on the other hand, characteristically extends a few branches onto the surfaces of muscle fibers near the horizontal myoseptum, in the territory of the RoP motoneuron. These branches probably form functional neuromuscular junctions. Later, after arrival of branches from the RoP motoneuron onto the same muscle fibers, the CaP branches are retracted. Branch retraction precedes breakup of the acetylcholine receptor clusters, occurs in the absence of the other motoneurons, and proceeds independently of neuronal or muscular activity.

We suggest that the few misplaced branches found in the zebrafish primary motor system represent the byproduct of the tightly regulated mechanisms that underlie cell-specific pathfinding, rather than the exuberant overproduction of branches and subsequent pruning that is observed during polyneuronal innervation in other vertebrate neuromuscular systems. Perhaps the primary motoneurons fulfill a developmental function, to pioneer the peripheral nerves, in addition to serving a role in behavior (37), and this developmental role is regulated by a set of mechanisms different than those that regulate the matching of populations of motoneurons to different sized muscles by activity-dependent synapse elimination.

ACKNOWLEDGMENTS

We thank D. Brumbley, J. Gleason and S. Posten for technical assistance. Supported by NIH NS21132, HD22486, and GM7257.

REFERENCES

1. Kimmel, C. B., and Westerfield, M. (1990). In "Signal and Sense" (G. M. Edelman, W. E. Gall, and M. W. Cowan, eds.), pp. 561–588. Wiley–Liss, New York.
2. Goodman, C. S., Raper, J. A., Ho, R. K., and Chang, S. (1982). Symp. Soc. Dev. Biol. 40, 275–316.
3. Bentley, D., and Keshishian, H. (1982). Science 218, 1082–1088.
4. Blair, S. B., and Palka, J. (1985). Trends Neurosci. 8, 284–288.
5. Ball, E. E., Ho, R. K., and Goodman, C. S. (1985). J. Neurosci. 5, 1808–1819.
6. Johansen, J., Halpern, M. E., and Keshishian, H. (1989). J. Neurosci. 9, 4318–4332.
7. Lance-Jones, C., and Landmesser, L. (1981). Proc. R. Soc. Lond. (biol.) 214, 19–52.
8. Westerfield, M., and Eisen, J. S. (1985). Dev. Biol. 104, 96–101.
9. Farel, P. B., and Bemelmans, S. E. (1985). J. Comp. Neurol. 238, 128–134.
10. Eisen, J. S., Myers, P. Z., and Westerfield, M. (1986). Nature 320, 269–271.
11. Myers, P. Z., Eisen, J. S., and Westerfield, M. (1986). J. Neurosci. 6, 2278–2289.
12. Westerfield, M., McMurray, J. V., and Eisen, J. S. (1986). J. Neurosci. 6, 2267–2277.
13. Eisen, J. S., Pike, S. H., and Romancier, B. (1990). J. Neurosci. 10, 34–43.
14. Eisen, J. S., Pike, S. H., and Debu, B. (1989). Neuron 2, 1097–1104.
15. Pike, S., and Eisen, J. S. (1990). J. Neurosci. 10, 44–49.

16. Westerfield, M., and Eisen, J. S. (1988). *Trends Neurosci.* **11**, 18–22.
17. Frost, D., and Westerfield, M. (1986). *Soc. Neurosci. Abstr.* **12**, 1114.
18. Westerfield, M. (1987). *J. Exp. Biol.* **132**, 161–175.
19. Van Essen, D. C. (1982). *In* "Neuronal Development" (N. C. Spitzer, ed.). Plenum Press, New York/London.
20. Oppenheim, R. W. (1981). *J. Neurosci.* **1**, 141–151.
21. Betz, W. J. (1987). *In* "The Vertebrate Neuromuscular Junction" (M. M. Salpeter, ed.), pp. 117–162. A. R. Liss, New York.
22. Purves, D., and Lichtman, J. W. (1980). *Science* **210**, 153–157.
23. Murphey, R. K. (1986). *J. Comp. Neurol.* **251**, 100–110.
24. Gao, W.-Q., and Macagno, E. R. (1987). *J. Neurobiol.* **18**, 295–313.
25. Oppenheim, R. W. (1981). *In* "Studies in Developmental Neurobiology: Essays in Honor of Viktor Hamburger" (W. M Cowan, ed.). Oxford Univ. Press.
26. Purves, D., and Lichtman, J. W. (1985). "Principles of Neural Development." Sinauer Associates Inc., Sunderland, MA.
27. Betz, W. J., Ribchester, R. R., and Ridge, R. M. A. P. (1990). *J. Neurobiol.* **21**, 1–17.
28. Liu, D. W., and Westerfield, M. (1990). *J. Neurosci.* **10**, 3947–3959.
29. Harris, W. A., Holt, C. E., and Bonhoeffer, F. (1987). *Development* **101**, 123–133.
30. O'Leary, D. D. M., and Terashima, T. (1988). *Neuron* **1**, 901–910.
31. Liu, D. W. and Westerfield, M. (1992). *J. Neurosci.* **12**, 1859–1866.
32. Westerfield, M., Liu, D. W., Kimmel, C. B., and Walker, C. (1990). *Neuron* **3**, 867–874.
33. Grunwald, D. J., Kimmel, C. B., Westerfield, M., Walker, C., and Streisinger, G. (1988). *Dev. Biol.* **126**, 115–128.
34. Rich, M. M., and Lichtman, J. W. (1989). *J. Neurosci.* **9**, 1781–1805.
35. Caudy, M., and Bentley, D. (1986). *J. Neurosci.* **6**, 1781–1795.
36. Myers, C. M., Whitington, P. M., and Ball, E. E. (1990). *Dev. Biol.* **137**, 194–206.
37. Liu, D. W., and Westerfield, M. (1988). *J. Physiol.* **403**, 73–89.

PART IV

GRADIENTS AND DIFFUSIBLE SIGNALS

9

The DVR Gene Family in Embryonic Development

KAREN M. LYONS* AND BRIGID L. M. HOGAN
Department of Cell Biology
Vanderbilt University School of Medicine
Nashville, Tennesee 37212
and
*Department of Cellular and Developmental Biology
Harvard University
Cambridge, Massachusetts 02138

INTRODUCTION

The DVR gene family encodes a group of highly conserved extracellular protein molecules that have been shown to mediate tissue interactions during embryonic development. Because the members of this family were discovered in different laboratories, using a variety of methods, the nomenclature has become somewhat confusing. One solution has been to rename them the DVR family (*Drosophila* decapentaplegic, *Xenopus* Vg-1-related), after the first members to be described (Padgett *et al.*, 1987; Weeks and Melton, 1988) (Table I). The DVR gene family includes the mammalian bone morphogenetic proteins (BMPs) -2, -3 (osteogenin), -4, and -5 (Wozney *et al.*, 1988; Luyten *et al.*, 1989; Celeste *et al.*, 1990), as well as murine Vgr-1 (BMP-6) (Lyons *et al.*, 1989a), and osteogenic protein 1 (BMP-7) (Ozkaynak *et al.*, 1990). The DVR family belongs to the much larger transforming growth factor-β (TGF-β) superfamily, which includes TGF-βs 1 through 5, as well as the activins, inhibins, Müllerian inhibiting substance, and GDF-1.

The members of the TGF-β superfamily, including the DVR gene products, are secreted proteins that are synthesized as prepropeptides with a highly conserved C-terminal region containing seven or nine cysteine residues. Dimerization, followed by proteolytic cleavage, results in a biologically active C-terminal dimer of approximately M_4

Cell–Cell Signaling in
Vertebrate Development

TABLE I
The DVR Family of TGF-β-Related Proteins[a]

Mammalian	Xenopus	Drosophila
	DVR-1/Vg1	
DVR-2/BMP-2/BMP-2a	DVR-2	DPP
DVR-4/BMP-4/BMP-2b	DVR-4	
DVR-3/osteogenin	DVR-3	
DVR-5/BMP-5	DVR-5	
DVR-6/BMP-6/Vgr-1	DVR-6	dVgr/60A
DVR-7/BMP-7/OP-1	DVR-7	
	DVR-8-14	

[a] Alternative names are also indicated (see text for references). Probable cognate proteins are listed on the same line. For example, *Drosophila* Vgr (dVgr/60A) is closely related to DVR-6 in mammals and *Xenopus*.

25,000–30,000 (Fig. 1). Because the receptors for DVR proteins have not been identified, the mechanism of signal transduction following binding of these dimers is unknown. However, the recent cloning of a high-affinity binding protein for activin, and its homology to serine–threonine kinase receptors from *Caenorhabditis elegans* and maize, suggests that similar receptors will be identified for DVR proteins (Mathews and Vale, 1991). TGF-βs have also been shown to bind with lower affinity to extracellular matrix components, suggesting that an important role for the extracellular matrix is to regulate the availability of these growth factors during embryonic development (Massagué, 1991).

Members of the DVR family have been shown by genetic and biochemical criteria to be involved in cell–cell interactions during development in a diverse group of species. The mammalian DVR-2,-3, and

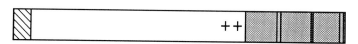

Fig. 1. Schematic representation of one polypeptide chain of a DVR protein. The hatched region represents a hydrophobic leader sequence, and the stippled area represents the highly conserved, mature region of the protein, which is about 100 amino acids long. The ++ symbol represents a cluster of basic residues at which proteolytic cleavage occurs. The seven vertical lines within the mature region show the relative positions of conserved cysteine residues, which form disulfide bonds between two polypeptide chains to form dimers. Potential glycosylation sites (not shown) are also present within the protein sequence.

-4 proteins are capable of inducing ectopic cartilage and bone formation when implanted subcutaneously in rats, suggesting an important
role in skeletal development (Wozney *et al.*, 1988; Hammonds *et al.*,
1991). Data from our laboratory, discussed below, are consistent with
this idea and suggest additional roles for certain DVR gene products
during mammalian development.

The most well-characterized DVR family member is the *Drosophila*
dpp gene product. Genetic analyses have demonstrated that *dpp* function is necessary at many stages of development. For example, DPP is
required for the specification of the dorsoventral axis, since embryos
homozygous for null alleles of *dpp* are completely ventralized (Gelbart, 1989). DPP is also necessary for normal development along the
proximodistal axis in imaginal disks, since certain *dpp* mutations affect
the morphology of adult structures derived from these disks (Gelbart,
1989). More recent work has focused on the role of the DPP protein in
the development of the midgut, which is divided by constrictions into
a number of segments, during larval development. This process involves inductive interactions between the mesenchyme of the midgut
and the underlying enidoderm. The production of antibodies specific
to DPP, coupled with genetic analyses, has revealed that DPP functions as an intercellular signaling molecule to mediate these inductive
interactions (Immerglück *et al.*, 1990; Reuter *et al.*, 1990; Panganiban
et al., 1990). Briefly, the expression of the homeotic gene *Ubx* in the
mesenchyme of parasegment 7 of the developing midgut results in the
expression of DPP. DPP then moves locally across the mesodermal
layer to the underlying endoderm, where it induces the expression of
the homeobox gene, *labial*. During midgut development, DPP functions both upstream and downstream of specific homeobox genes,
whose expression is required for the development of the midgut. DPP
thus serves as an intercellular signal between two tissue layers, mediating a transfer of positional information from one layer (the mesenchyme) to another (the endoderm). The action of DPP as a signaling
molecule mediating inductive interactions in the developing larval
midgut may serve as a paradigm for the roles of other DVR family
members during vertebrate embryogenesis.

One of the most intensively studied inductive interactions during
development is the initial establishment of mesoderm in *Xenopus*. A
role for TGF-β-related proteins, including DVR family members, in
mesoderm induction is supported by data from a number of laboratories. *Xenopus Vg-1*, a founding member of the DVR family, was originally identified as a mRNA localized to the vegetal pole of *Xenopus*
oocytes (Weeks and Melton, 1988). This observation raised the possi-

bility that the Vg-1 protein might serve as a signaling molecule during the inductive interaction that takes place between cells in the animal and vegetal poles, resulting in the specification of mesoderm in the equatorial marginal zone. Whether Vg-1 has such activity is not yet known. However, a number of experiments point to activin, a TGF-β-related molecule, in the specification of dorsal mesoderm (Smith *et al.*, 1990; van den Eijinden-Van Raaij *et al.*, 1990; Thomsen *et al.*, 1990). Mesoderm induction in *Xenopus* most likely involves the combinatorial action of a number of growth factors. For example, basic fibroblast growth factor (bFGF) (Slack *et al.*, 1987), the Wnt-related gene product, *wnt*-8 (Christian *et al.*, 1991), and DVR-4 (Köster *et al.*, 1991) have all been implicated in the specification of mesodermal cell fate. Experiments from our laboratory also imply a role for DVR-4 in the specification of ventral mesoderm during *Xenopus* development (Jones *et al.*, 1992). The localization of DVR-4 transcripts in the newly formed ventral, posterior mesoderm of the mouse embryo at 8.5 days pc is consistent with the possibility of a similar function during mammalian development (Jones *et al.*, 1991).

RESULTS AND DISCUSSION

Given the evidence that DVR family members participate in inductive interactions during *Drosophila* and *Xenopus* development, it is reasonable to expect similar roles for DVR proteins during mammalian embryogenesis. One prediction of this hypothesis is that DVR genes should be expressed in temporally and spatially restricted patterns in tissues where development proceeds through a series of epithelial–mesenchymal interactions. We examined the patterns of expression of DVR-2,-4, and -6 RNAs during mouse development. As it turns out, these genes have cognates in *Drosophila* (Table I), raising the possibility of some conserved mechanisms of action.

Detailed *in situ* hybridization studies have revealed expression at all stages of development and in derivatives of all three germ layers (Lyons *et al.*, 1989b, 1990; Jones *et al.*, 1991). DVR-2, -4, and -6 are expressed in many of the same developing tissues (for example, heart, limb buds, cartilage and bone, whisker follicles, and central nervous system), but with temporally and spatially distinct patterns, suggesting that the coordinated expression of multiple members of the gene family is required to orchestrate the development of complex organs (Lyons *et al.*, 1990). Furthermore, these genes are often expressed in both epithelial and mesenchymal components of tissues where development is known to involve reciprocal inductive interactions. This concept is illustrated in the developing limb bud, whisker follicle, and heart.

Expression of DVR-2 and DVR-4 in Limb Buds

Classic transplantation studies have shown that the establishment of the proximodistal and dorsoventral axes of the limb bud involves inductive interactions between the outer ectodermal layer and the inner mesodermal layer (Tickle, 1980; Hinchcliffe and Johnson, 1980; Smith *et al.*, 1989). Development along the proximodistal axis is thought to be controlled by signals arising in the slightly thickened epithelium, the apical ectodermal ridge (AER), at the tip of the limb bud. These signals influence cells in the underlying progress zone so that they continue to divide and do not differentiate. Consistent with this view, removal of the AER leads to limb truncations. Similarly, there is some evidence from transplantation studies suggesting that ventral ectoderm plays a role in patterning the underlying mesenchyme along the dorsoventral axis (MacCabe *et al.*, 1974; Patou, 1977).

A role for DVR-2 and DVR-4 in patterning in the limb bud is suggested by *in situ* hybridization analyses (Lyons *et al.*, 1990; Jones *et al.*, 1991). At the earliest stage examined, 9.5 days pc, the limb is a discrete outgrowth from the body wall, but the AER is not yet distinct in the mouse. DVR-2 transcripts are localized to the thickened ventral epithelium, raising the possibility of a role in dorsoventral patterning (Figs. 2A and 2B). By 10.5 days pc, the AER is distinguishable, and transcripts for both DVR-2 and DVR-4 are localized in this structure (Figs. 2C–2F). At this stage of development, DVR-4 RNA is also found within the underlying mesenchyme (Figs. 2E and 2F). It is tempting to speculate that DVR-2 and DVR-4 produced in the AER may diffuse to the underlying progress zone, where these proteins influence the pro-

---

Fig. 2. Expression of DVR-2 and DVR-4 in developing limb buds. (A) Bright-field and (B) dark-ground images of a section through a forelimb bud from a 9.5-day pc embryo hybridized to a DVR-2 probe. Transcripts are present in the thickened ventral epithelium. (C) Bright-field and (D) dark-ground photomicrographs of a section through a 10.5-day pc embryo, showing localization of DVR-2 RNA to the AER (arrow). (E) Bright-field and (F) dark-ground images of a section though a 10.5-day pc limb bud showing hybridization to the DVR-4 probe in both the AER and the underlying mesenchyme. ve, ventral epithelium; aer, apical ectodermal ridge; nt, neural tube; da, dorsal aorta; g, gut.

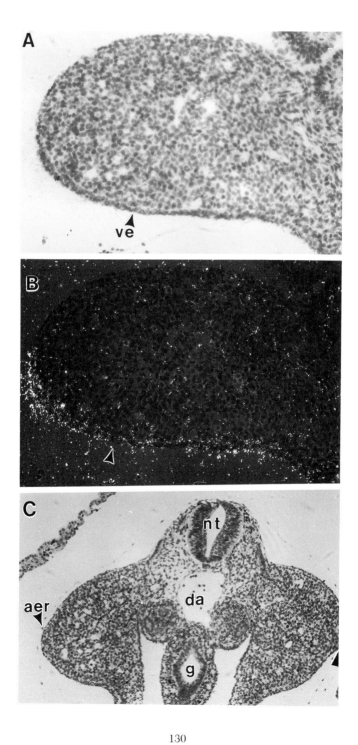

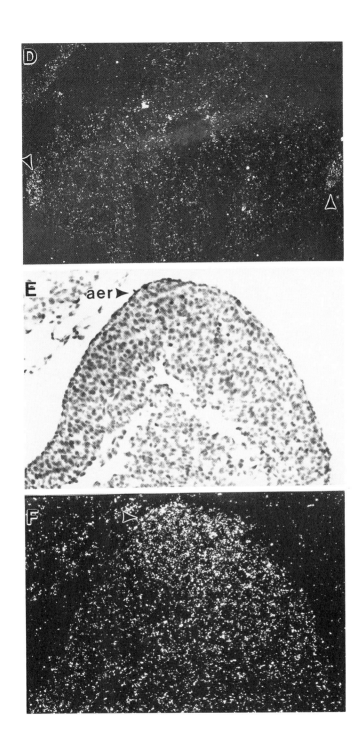

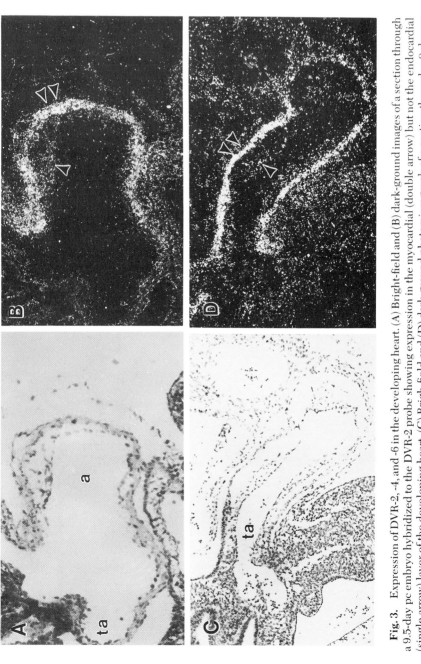

Fig. 3. Expression of DVR-2, -4, and -6 in the developing heart. (A) Bright-field and (B) dark-ground images of a section through a 9.5-day pc embryo hybridized to the DVR-2 probe showing expression in the myocardial (double arrow) but not the endocardial (single arrow) layer of the developing heart. (C) Brigh-field and (D) dark-ground photomicrograph of a section through a 9-day pc embryo showing expression of DVR-4 RNA within the myocardial (double arrow) layer but not the endocardial (single arrow) layer of the truncus arteriosus (ta).

liferation and fate of the mesenchymal cells. These proteins may additionally have an effect on the establishment and maintenance of the AER.

Information is rapidly accumulating about the expression in limb buds of putative patterning molecules, such as homeobox proteins, and retinoic acid receptors, as well as potential intercellular signaling molecules, such as the *wnt* family members (Dollé *et al.*, 1989; Nohno *et al.*, 1991; Izpisúa-Belmonte *et al.*, 1991; Davis *et al.*, 1991; Gavin *et al.*, 1990). It is tempting to speculate that the DVR-2 and DVR-4 gene products function as intercellular signaling molecules, regulating the expression of homeobox genes in the limb, in a manner analogous to that of DPP during larval midgut development in *Drosophila*. One prediction of this hypothesis is that altering the patterns of expression of DVR-2 and DVR-4, either transgenically or by implanting protein, should alter the patterns of expression of specific homeobox genes.

DVR Gene Expression in the Developing Heart

The development of the vertebrate heart also involves reciprocal inductive interactions. The heart develops from a simple tubular structure, composed of two concentric epithelial layers separated by a layer of extracellular matrix material. Beginning at around 9.5 days pc, dramatic morphogenetic movements take place, resulting in the formation of chambers separated by septa and valves. The formation of the endocardial cushions, which give rise to the septa, is initiated by the localized differentiation of endocardial cells into fibroblast-like cells, which migrate into the underlying extracellular matrix. At 9.5 days pc, DVR-2 and DVR-4 RNAs are present throughout the myocardial layer of the developing heart (Figs. 3A–3D). By 10.5 days pc, both DVR-2 and DVR-4 transcripts are localized to the myocardial cells underlying the developing endocardial cushions (Lyons *et al.*, 1990; Jones *et al.*, 1991). DVR-2 and DVR-4 are not the only members of the TGF-β family expressed in the endocardial cushions; we have observed DVR-6 transcripts in the mesenchymal cells lying between the endocardial and myocardial layers, and TGF-β_1 and -β_2 RNAs are present in the endocardial and mesenchymal layers, respectively (Lehnert and Akhurst, 1988; Pelton *et al.*, 1990; Millan *et al.*, 1991). The expression of multiple TGF-β-related genes within tissues whose development proceeds through inductive interactions is a common theme and raises the possibility that their coordinated expression is required for the development of these organs. These patterns of expression also raise the possibility that DVR proteins act as intercellular signals to regulate the expression of other DVR family members.

In vitro studies support a role for TGF-β-related proteins during endocardial cushion formation. The epithelial–mesenchymal transformation of endocardial cells into fibroblast-like cells during cushion formation is induced by signals found in the extracellular matrix of the myocardial cells (Runyon and Markwald, 1983; Mjaatvedt and Markwald, 1989). TGF-β_1 and -β_2 enhance the migration of endothelial cells *in vitro*, but these proteins are insufficient to induce the formation of migratory mesenchymal cells from endothelium. An additional signal from the myocardium is required (Potts and Runyon, 1989). Our localization of DVR-2 and DVR-4 RNAs in the myocardium at the time of active formation of the cushions raises the possibility that DVR-2 and DVR-4 proteins may be the necessary signal. One test of this hypothesis is to examine the effects of additional DVR-2 or DVR-4 protein, either *in vitro* on endocardial cells or *in vivo* by generating transgenic mice which overexpress these proteins in the heart.

Evolution of the DVR Gene Family

Only two members of the DVR family have been identified in *Drosophila*. One of these is *dpp*, which, as discussed above, plays multiple roles during development. The function of the second gene, named dVgr and 60A is as yet unknown (John Doctor and Michael Hoffmann, personal communication). In contrast, multiple members of the DVR family are found in vertebrates (Table I). These genes can be arranged into two distinct groups, based on amino acid homologies within the conserved C-terminal regions. DVR-2 and DVR-4 are closely related to each other, and are likely the vertebrate cognates of *dpp*. DVR-5, -6, and -7 form a second group and are closely related to dVgr. Thus the DVR family can presently be divided into two distinct classes, and this division is conserved in vertebrates and *Drosophila*. We have employed a polymerase chain reaction (PCR) approach to identify three additional DVR family members in the mouse. (K. Lyons, and G. Thomsen, unpublished data). Additional sequence data are necessary before their relationship to published DVR genes is known.

The conservation of distinct groups of DVR genes across a wide phylogenetic distance suggests distinct functions for the members of each class. A thorough genetic analysis, involving misexpression in transgenic mice, and gene targeting to generate null alleles will be necessary to understand the function of each DVR family member. Other remaining challenges are to unravel the significance and molecular basis for the coordinated expression of multiple DVR family members, in certain developing systems, and mechanisms of signal transduction. Fortunately, many of the necessary tools are now available to approach these and related questions.

REFERENCES

Celeste, A. J., Iannazzi, J. A., Taylor, R. C, Hewick, R. M., Rosen, V., Wang, E. A., and Wozney, J. M. (1990). Identification of transforming growth factor β family members present in bone-inductive protein purified from bovine bone. *Proc. Natl. Acad. Sci. U.S.A.* **87**, 9843–9847.

Christian, J. L., McMahon, J. A., McMahon, A. P., and Moon, R. T. (1991). *Xwnt-8*, a *Xenopus Wnt-1/int-1*/related gene responsive to mesoderm-inducing growth factors, may play a role in ventral mesodermal patterning during embryogenesis. *Development* **111**, 1045–1055.

Davis, C. A., Holmyard, D. P., Millen, K. J., and Joyner, A. L. (1991). Examining pattern formation in mouse, chicken and frog embryos with an *En*-specific antiserum. *Development* **111**, 287–298.

Dollé, P., Ruberte, E., Kastner, P., Petkovich, P., Stoner, C. M., Gudas, L. J., and Chambon, P. (1989). Differential expression of genes encoding α, β, and γ retinoic acid receptors and CRABP in the developing limbs of the mouse. *Nature* **342**, 702–705.

Gavin, B. J., McMahon, J. A., and McMahon, A. P. (1990). Expression of multiple novel *Wnt-1/int-1*-related genes during fetal and adult mouse development. *Genes Dev.* **4**, 2319–2332.

Gelbart, W. M. (1989). The *decapentaplegic* gene: A TGF-β homologue controlling pattern formation in *Drosophila*. *Development Suppl.* **107**, 65–74.

Hammonds, R. G., Schall, R., Dudley, A., Berkemeier, L., Lai, C., Lee, J., Cunningham, N., Reddi, A. H., Wood, W. I., and Mason, A. J. (1991). Bone inducing activity of mature BMP-2b produced from a hybrid Bmp-2a/2b precursor. *Mol. Endocrinol.* **5**, 149–155.

Hinchcliffe, J. R., and Johnson, D. R. (1980). "The Development of the Vertebrate Limb." Clarendon, Oxford.

Immerglück, K., Lawrence, P. A., and Bienz, M. (1990). Induction across germ layers in Drosophila mediated by a genetic cascade. *Cell* **62**, 261–268.

Izpisúa-Belmonte, J. C., Tickle, C., Dollé, P., Wolpert, L., and Duboule, D. (1991). Expression of the homeobox Hox-4 genes and the specification of position in chick wing development. *Nature* **350**, 585–589.

Jones, C. M., Lyons, K. M., and Hogan, B. L. M. (1991). Involvement of *Bone Morphogenetic Protein-4(BMP-4)* and *Vgr-1* in morphogenesis and neurogenesis in the mouse. *Development* **111**, 531–542.

Jones, C. M., Lyons, K. M., Lapan, P. M., Wright, C. V. E., and Hogan, B. L. M. (1992). DVR-4 (bone morphogenetic protein-4) as a posterior-ventralizing factor in *Xenopus* mesoderm induction. *Development* **115**, 639–647.

Köster, M., Plessow, S., Clement, J. H., Lorenz, A., Tiedeman, H., and Knöchel, W. (1991). Bond morphogenetic protein 4 (BMP-4), a member of the TGF-β family, in early embryos of *Xenopus laevis:* Analysis of mesoderm inducing activity. *Mech. Dev.* **33**, 191–200.

Lehnert, S. A., and Akhurst, R. J. (1988). Embryonic expression pattern of TGF beta type-1 RNA suggests both paracrine and autocrine mechanisms of action. *Development* **104**, 263–273.

Luyten, F. P., Cunningham, N. S., Ma, S., Muthukumaran, N., Hammonds, R. G., Nevins, W. B., Wood, W. I., and Reddi, H. (1989). Purification and partial amino acid sequence of osteogenin, a protein initiating bone differentiation. *J. Biol. chem.* **264**, 13377–13380.

Lyons, K. M., Graycar, J. L., Lee, A., Hashmi, S., Lindquist, P. B., Chen, E. Y., Hogan, B. L. M., and Derynck, R. (1989a). Vgr-1, a mammalian gene related to Xenopus Vg-1, is a member of the transforming growth factor β gene superfamily. *Proc. Natl. Acad. Sci. U.S.A.* **86**, 4554–4558.

Lyons, K. M., Pelton, R. W., and Hogan, B. L. M. (1989b). Patterns of expression of murine Vgr-1 and BMP-2a suggest that TGFβ-like genes coordinately regulate aspects of embryonic development. *Genes Dev.* **3**, 1657–1668.

Lyons, K. M., Pelton, R. W., and Hogan, B. L. M. (1990). Organogenesis and pattern formation in the mouse: RNA distribution pattern suggest a role for *Bone Morphogenetic Protein-2A (BMP-2A)*. *Development* **109**, 833—844.

MacCabe, J. A., Errick, J., and Saunders, J. W. (1974). Ectodermal control of dorsoventral axis in the leg bud on the chick embryo. *Dev. Biol.* **39**, 69–82.

Massagué, J. (1991). A helping hand from proteoglycans. *Curr. Biol.* **1**, 117–119.

Mathews, L. S., and Vale, W. W. (1991). Expression cloning of an activin receptor, a predicted transmembrane serine kinase. *Cell* **65**, 973–982.

Millan, F. A., Denhez, F., Kondaiah, P., and Akhurst, R. J. (1991). Embryonic gene expression patterns of TGF β1, β2 and β3 suggest different developmental functions *in vivo*. *Development* **111**, 131–144.

Mjaatvedt, C. H., and Markwald, R. R. (1989). Induction of an epithelial-mesenchymal transition by an *in vitro* adheron-like complex. *Dev. Biol.* **136**, 118–128.

Nohno, T., Noji, S., Koyama, E., Ohyoama, K., Myokai, F., Kuroiwa, A., Saito, T., and Taniguchi, S. (1991). Involvement of the Chox-4 chicken homeobox genes in determination of anteroposterior axial polarity during limb development. *Cell* **64**, 1197–1205.

Ozkaynak, E., Rueger, D. C., Drier, E. A., Corbett, C., Ridge, R. J., Sampath, T. K., and Opperman, H. (1990). OP-1 cDNA encodes an osteogenic protein in the TGF-β family. *EMBO J.* **9**, 2085–2093.

Padgett, R., St. Johnston, R. D., and Gelbart, W. M. (1987). A transcript from a *Drosphila* pattern gene predicts a protein homologous to the transforming growth factor-beta family. *Nature* **325**, 81–84.

Panganiban, G. E. F., Reuter, R., Scott, M. P., and Hoffmann, F. M. (1990). A *Drosophila* growth factor homolog, *decapentaplegic*, regulates homeotic gene expression within and across germ layers during midgut morphogenesis. *Development* **110**, 1041–1050.

Patou, M. P. (1977). Dorso-ventral axis determination of chick limb bud development. *In* "Vertebrate Limb and Somite Morphogenesis" (D. A. Ede, J. R. Hinchcliffe, and M. Balls, eds.), pp. 257–266. Cambridge Univ. Press.

Pelton, R. W., Dickinson, M. E., Moses, H. L., and Hogan, B. L. M. (1990). *In situ* hybridization analysis of TGFβ3 RNA expression during mouse development: Comparative studies with TGFβ1 and β2. *Development* **110**, 609–620.

Potts, J. D., and Runyon, R. B. (1989). Epithelial-mesenchymal cell transformation in the embryonic heart can be mediated, in part, by transforming growth factor β. *Dev. Biol.* **134**, 392–401.

Reuter, R., Panganiban, G. E. F., Hoffmann, F. M., and Scott, M. (1990). Homeotic genes regulate the spatial expression of putative growth factors in the visceral mesoderm of *Drosophila* embryos. *Development* **110**, 1031–1040.

Runyon, R. B., and Markwald, R. R. (1983). Invasion of mesenchyme into three-dimensional collagen gels: A regional and temporal analysis of interaction in embryonic heart tissue. *Dev. Biol.* **85**, 108–114.

Slack, J. M. W., Darlington, B. G., Heath, J. K., and Goodsave, S. F. (1987). Mesoderm induction in early Xenopus embryos by heparin-binding growth factors. *Nature* **326**, 197–200.

Smith, S. M., Pang, K., Sundin, O., Wedden, S. E., Thaller, C., and Eichele, G. (1989). Molecular approaches to vertebrate limb morphogenesis. *Development* **107** (Suppl.), 121–131.

Smith, J. C., Price, B. M. J., van Nimmen, K., and Huylebroeck, D. (1990). Identification of a potent Xenopus mesoderm-inducing factor as a homologue of activin A. *Nature* **345**, 729–731.

Thomsen, G., Woolf, T., Whitman, M., Sokol, S., Vaughan, J., Vale, W., and Melton, D. A. (1990). Activins are expressed early in Xenopus embryogeneis and can induce axial mesoderm and anterior structures. *Cell* **63**, 485–493.

Tickle, C. (1980). *In* ''Development in Mammals'' (M. H. Johnson, ed.), Vol. 4, pp. 101–136. Elsevier/North-Holland Biochemical Press, Amsterdam.

van den Eijnden-Van Raaij, A. J. M., van Zoelent, E. J. J., van Nimmen, K., Koster, C. H., Snoek, G. T., Durston, A. J., and Huylebroeck, D. (1990). Activin-like factor from a Xenopus laevis cell line responsible for mesoderm induction. *Nature* **345**, 732–734.

Weeks, D. L., and Melton, D. A. (1988). A maternal mRNA localized to the vegetal hemisphere in Xenopus eggs codes for a growth factor related to TGF-β. *Cell* **51**, 861–867.

Wozney, J. M., Rosen, V., Celeste, A. J., Mitsock, L. M., Whitters, M. J., Kriz, R. W., Hewick, R. M., and Wang, E. A. (1988). Novel regulators of bone formation: Molecular clones and activities. *Science* **242**, 1528–1534.

10

Control of Neural Cell Identity and Pattern by Notochord and Floor Plate Signals

THOMAS M. JESSELL* AND JANE DODD†
*Department of Biochemistry and Molecular Biophysics
Howard Hughes Medical Institute
Center for Neurobiology and Behavior
Columbia University
New York, New York 10032
and
† Department of Physiology and Cellular Biophysics
Center for Neurobiology and Behavior
Columbia University
New York, New York 10032

INTRODUCTION

One of the central issues in vertebrate neural development is to define the cellular interactions that control the identity and patterning of specific neural cell types. The establishment of cell pattern along the anteroposterior (A–P) axis of the neural plate is evident soon after the induction of neural ectoderm. Anterior regions of the neural plate give rise to the forebrain and midbrain and posterior regions to the hindbrain and spinal cord. The pattern of cell differentiation along the dorsoventral (D–V) axis occur later, as the neural plate folds to form the neural tube (Roach, 1945; Jacobson, 1964) (Fig. 1). In anterior regions of the neural plate, small differences in the position of a cell along the A–P axis result in a marked difference in its fate (Puelles et al., 1987). In contrast, in the spinal cord the identity and pattern of cells is similar at different A–P levels (Hollyday and Hamburger, 1977; Eisen, 1991).

Cell–Cell Signaling in
Vertebrate Development

Neural Plate   Neural Fold   Neural Tube   Spinal Cord

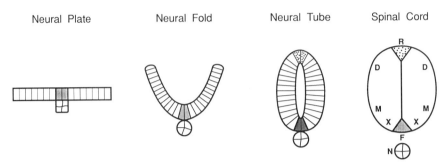

Fig. 1. Establishment of the dorsoventral axis in the developing spinal cord. The four diagrams show successive stages in the development of the neural tube and spinal cord. The neural plate consists initially of a simple columnar epithelium. Cells at the midline of the neural plate are contacted directly by axial mesoderm cells of the notochord. More lateral regions of the neural plate overlie the segmental plate mesoderm (not shown). During neurulation, the neural plate buckles at its midline to form the neural folds. Contact between the midline of the neural plate and the notochord is maintained at this stage. The neural tube is formed when the dorsal tips of the neural folds fuse. Cells in the region of fusion form a specialized group of dorsal midline cells, the roof plate. Cells at the ventral midline of the neural tube retain proximity to the notochord and differentiate into the floor plate. After neural tube closure neuroepithelial cells continue to proliferate and eventually differentiate into defined classes of neurons at different dorsoventral positions within the spinal cord. For example, commissural and other classes of dorsal neurons (D) differentiate near to the roof plate (R), and motor (M) neurons differentiate ventrally near to the floor plate (F), which by this time is no longer in contact with the notochord (N). Other classes of neurons (X) differentiate in the region of neural epithelium that intervenes between the floor plate and motor neurons. For further details see Schoenwolf and Smith (1990) and Yamada *et al.* (1991).

Thus, neural cell identity in the developing spinal cord is likely to be dependent primarily on the position that cells occupy along the D–V axis of the neural tube.

During spinal cord development, the first neural cell group to differentiate is located at the midline of the neural plate and gives rise to the floor plate at the ventral midline of the neural tube (Baker, 1927; Kingsbury, 1930; Schoenwolf and Smith, 1990) (Fig. 1). Several lines of evidence indicate that the floor plate has highly specialized properties. The floor plate is the source of a polarizing signal that causes mirror-image digit duplications along the A–P axis of the developing chick limb (Wagner *et al.*, 1990), mimicking the effect of the limb polarizing region and of retinoic acid (Tickle *et al.*, 1975, 1982). The floor plate also appears to contribute to the pathfinding of axons in the developing spinal cord (Jessell *et al.*, 1989). The growth of the axons of spinal commissural neurons toward the ventral midline of the embryonic spinal cord appears to be guided by a diffusible chemoattractant released by the floor plate (Tessier-Lavigne *et al.*, 1988; Placzek *et al.*, 1990a, b). In addition, the floor plate appears to contribute to changes in axonal trajectory at the midline (Bovolenta and Dodd, 1990, 1991; Kuwada *et al.*, 1990a, b; Kuwada and Hatta, 1990; Hatta *et al.*, 1990; Yaginuma *et al.*, 1991).

The midline location of the floor plate, together with its signaling properties, raises the possibility that the floor plate regulates the identity of neurons as well as the guidance of axons in the developing spinal cord. This chapter reviews experiments which suggest that local inductive signals from the notochord control the differentiated properties of floor plate cells and that signals from the floor plate and notochord regulate the differentiation of neurons along the D–V axis of the neural tube.

INDUCTION OF THE FLOOR PLATE BY THE NOTOCHORD

Early studies on the control of floor plate differentiation examined the morphology of the spinal cord in chick embryos with duplicated notochords (Watterson *et al.*, 1955). Spinal cord cells with a characteristic floor plate morphology appeared adjacent to both notochords, suggesting that the morphological features of the floor plate depend on the notochord. Experimental studies have provided support for this idea. Grafting a notochord next to the lateral part of the neural tube results in the appearance of cells with floor plate-like morphology in the adjacent neural epithelium (van Straaten *et al.*, 1985a,b, 1988; Smith and Schoenwolf, 1989). In addition, in embryos in which the notochord has been deleted, wedge-shaped cells are absent from the midline of the spinal cord (Wolff, 1936; Grabowski, 1956; van Straaten *et al.*, 1987; Kuwada and Hatta, 1990; Hatta *et al.*, 1990, Bovolenta and Dodd, 1991).

To determine more directly whether floor plate differentiation depends on signals that derive from the notochord, surface antigens expressed selectively by the floor plate have been used to examine cell pattern in embryos with notochord grafts. Several different floor plate antigens, FP1, FP2, and SC1, are induced in neuroepithelial cells adjacent to notochord grafts, providing evidence that biochemical markers of floor plate differentiation are induced by the notochord (Yamada *et al.*, 1991). In addition, the floor plate-specific chemoattractant involved in commissural axon guidance (Tessier-Lavigne *et al.*, 1988; Placzek *et al.*, 1990a) provides an assay with which to examine whether the notochord can induce functional properties characteristic of the floor plate. Explants from the region of spinal cord adjacent to a grafted notochord express chemoattractant activity whereas neural tissue from the contralateral, unoperated side does not (Placzek *et al.*, 1990c). Thus, the notochord can induce chemoattractant activity in regions of the chick neural tube that normally lack this activity.

To determine whether the notochord is required for the differentiation of the floor plate during normal development, the notochord was removed from chick embryos at stage 10 and expression of antigenic and functional properties of the floor plate examined after 24–48 hr. In the region of spinal cord lacking the notochord, cells at the ventral midline of the spinal cord do not express floor plate-specific antigens or chemoattractant activity (Placzek *et al.*, 1990c; Yamada *et al.*, 1991). These results provide evidence that the presence of the notochord is essential for the differentiation of the floor plate.

These *in vivo* studies, however, do not exclude the possibility that signals from other tissues are also required for floor plate differentiation. To test whether signals from the notochord are sufficient for induction of a floor plate, chick notochord tissue has been placed *in vitro* together with lateral regions of the neural plate (Placzek *et al.*, 1990c). Explants of the neural plate do not express chemotropic activity when grown in isolation, but when grown in direct contact with the notochord produce significant chemotropic activity. Moreover, when the notochord and lateral neural plate are placed close together, but not in contact, chemotropic activity is not detected.

These results suggest that signals from the notochord are sufficient to induce a floor plate in competent neural plate tissue. Moreover, proximity or direct contact between notochord cells and the neural plate appears to be necessary for induction of the floor plate *in vitro*. Indeed, during the development of the neural tube only those cells contacted by the notochord exhibit the shape changes and express the surface antigens characteristic of early floor plate differentiation (Schoenwolf and Smith, 1990; Yamada *et al.*, 1991). The contact dependence of floor plate induction could reflect the membrane anchoring of an inductive ligand, the requirement for a high concentration of a diffusible factor, or the sequestration of a diffusible signal in the extracellular matrix that surrounds the notochord.

Studies in other vertebrate species have indirectly provided evidence for the induction of the floor plate by the notochord. Ultraviolet treatment of fertilized *Xenopus* eggs can lead to the development of tadpoles lacking a notochord. The neural tube of these embryos lacks ventral midline cells with characteristics of the floor plate (Clarke *et al.*, 1991). In addition, analysis of neural differentiation in the embryonic mouse mutants *Danforth's short tail* and *truncate* in which the notochord fails to differentiate or undergoes early degeneration has shown that the caudal region of the spinal cord lacks a morphogenically and antigenically defined floor plate (Theiler, 1959; Bovolenta and Dodd, 1991). However, at more rostral regions of the spinal cord of *Danforth's short tail* embryos the floor plate is present even though the notochord is absent (Bovolenta and Dodd, 1991). This may result from the persistence of primitive notochord cells for a time sufficient to induce a floor plate in anterior regions.

More generally, these results suggest that mesodermally derived inductive signals have sequential and progressively more refined roles in regulating neural differentiation. During gastrulation, these signals appear to control the initial induction of the neural plate and its early regionalization along the A–P axis (Mangold, 1933; Spemann, 1938; Dixon and Kintner, 1989; Ruiz i Altaba, 1990). Later, during neural tube development, the axial mesoderm appears to determine the identity of floor plate cells.

SIGNALS FROM THE FLOOR PLATE AND NOTOCHORD CONTROL CELL PATTERN IN THE VENTRAL HALF OF THE NEURAL TUBE

The ability of the floor plate to polarize the developing chick limb (Wagner *et al.*, 1990), together with its midline location within the neural tube, raised the possibility that the floor plate may function together with the notochord to control the pattern of cell differentiation along the D–V axis of the neural tube (Jessell *et al.*, 1989). To determine whether the fate of neuroepithelial precursors within the chick neural tube depends on their position with respect to the floor plate and notochord a series of antigens was identified that is selectively expressed by classes of cells that occupy different positions along the D–V axis of the spinal cord and hindbrain (Yamada *et al.*, 1991).

The cell surface glycoprotein SC1 has been used as a marker to determine whether the notochord and floor plate can regulate the differentiation of motor neurons. Labeling of normal embryos with monoclonal antibody (MAb) SC1 reveals that motor neurons differentiate in the ventral spinal cord near, but not immediately adjacent to the floor plate (Fig. 2a). In embryos with dorsal notochord grafts, SC1 expression reveals the presence of two ectopic motor columns that differentiate in a bilaterally symmetric manner around the newly induced floor plate (Fig. 2b) (Yamada *et al.*, 1991). When notochord grafts are placed lateral to the neural tube, midway between the roof plate and floor plate, an ectopic floor plate is again detected adjacent to the graft. In addition, an ectopic column of motor neurons develops in the dorsal spinal cord on the side of the graft, separated from the induced floor plate by a region of unlabeled cells (Fig. 2b).

Strikingly, floor plate grafts produce changes in the pattern of cell differentiation along the D–V axis of the spinal cord that are indistinguishable from those produced by notochord grafts (Fig. 2c) (Yamada *et al.*, 1991). Grafting a floor plate dorsally results in the presence of a dorsal floor plate and two additional motor columns that differentiate in a bilaterally symmetric position with respect to the grafted and induced floor plate. Thus, the pattern of cell differentiation in the neural tube can also be modified by signals that originate from the floor plate.

To examine whether the notochord and floor plate are required to regulate the pattern of cell differentiation in the neural tube, the development of the floor plate can be prevented by removing a segment of

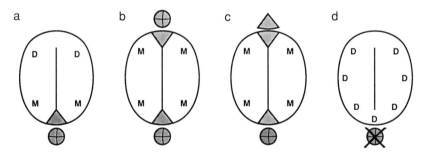

Fig. 2. Diagram summarizing the results obtained from experiments in chick embryos in which a notochord or floor plate is grafted to the dorsal midline of the neural tube and in which the notochord is removed before neural tube closure. (a) Neural organization of the spinal cord showing the ventral location of motor neurons (M) and the dorsal location of sensory relay neurons (D). (b) Dorsal grafts of a notochord result in the induction of a floor plate at the dorsal midline and in the induction of ectopic dorsal motor neurons. (c) Dorsal grafts of a floor plate also result in the induction of a floor plate at the dorsal midline and in the induction of ectopic dorsal motor neurons. (d) Removal of the notochord results in the elimination of the floor plate and motor neurons and the expression of dorsal cell types (D) in the ventral region of the spinal cord. For details see Yamada *et al.*, 1991; Placzek *et al.*, 1990c.

notochord underlying the caudal ventral midline of stage 10–12 embryos (Placzek *et al.*, 1990c; Yamada *et al.*, 1991). At distances greater than 100 μm from the end of the floor plate, motor neurons are absent. These results provide evidence that motor neuron differentiation is dependent on the floor plate or notochord.

The influence of the notochord and other mesodermal tissues on the differentiation of ventral neuroblasts in the chick neural tube has also been examined in other studies. Inversion of the neural tube along its D–V axis at stage 11–12 has been reported to result in the appearance of ventral characteristics in the formerly dorsal spinal cord, although specific cell types were not identified (Steding, 1962). Similarly, notochord grafts placed next to the neural tube have been shown to lead to the proliferation of ventral neuroblasts (van Straaten *et al.*, 1985b, 1988).

In chick spinal cord, motor neurons are separated from the floor plate by a region of intervening cells, from which a group of longitudinal interneurons (PL cells) differentiate (Yaginuma *et al.*, 1990). Antigenic markers that identify PL cells or other cells in region X in the spinal cord have not been defined. In the hindbrain, however, a subset of cells in region X differentiate into serotonergic neurons (Wallace, 1985). Grafting a piece of notochord adjacent to the neural tube at the level of the hindbrain induces serotonergic neurons in the region immediately adjacent to the ectopic floor plate (Yamada *et al.*, 1991). These findings show that cells characteristic of region X can also be induced by midline-derived signals and that they occupy a position between the ectopic floor plate cells and motor neurons.

The ability of the floor plate to mimic the inductive actions of the notochord indicates that the floor plate has homeogenetic inductive properties. Similar conclusions have been reached from an analysis of zebrafish *cyclops* mutant embryos (Hatta *et al.*, 1991). The homeogenetic inducing activity of the floor plate requires a mechanism for preventing the lateral propagation of this signal throughout the neuroepithelium, which would convert the entire neural tube into a floor plate. One possibility is that the notochord and/or floor plate provide long-range signals that induce more lateral neural plate cells to differentiate into ventral neural cell types such as cells in region X and motor neurons. The specification of these ventral cell types may occur over the same time period, or shortly after the time that the midline cells acquire primary floor plate properties. Thus, by the time that floor plate cells have acquired homeogenetic inducing properties few if any of cells remain competent for additional recruitment to a floor plate pathway of differentiation.

The cellular events involved in the differentiation of motor neurons in response to notochord and floor plate-derived signals remain unclear. Motor neuroblasts derive from the ventricular zone in the ventral region of the neural tube and migrate laterally to form a discrete cluster of motoneurons beginning at stage 15 (Langman and Haden, 1970; Hollyday and Hamburger, 1977). There is an increase in the total number of motoneurons after notochord and floor plate grafts (Yamada et al., 1991) which excludes the possibility that the appearance of ectopic motor neurons results from the redistribution of a fixed number of motor neuron precursors. Moreover, there is very little cell death in the lateral neural tube before stage 15 (Homma et al., 1990) which makes it unlikely that the appearance of ectopic motoneurons results from the rescue of a population of committed motoneuron progenitors. It seems more likely that the differentiation of motoneurons in ectopic regions of the spinal cord in response to notochord or floor plate grafts results from a change in the fate of pluripotent neuroepithelial progenitors. In support of this, lineage analysis of retrovirally marked cells in the chick neural tube indicates that motoneurons derive from progenitors that give rise to other classes of neurons and also to glial cells (Leber et al., 1990). Such multipotential progenitors may be distributed throughout the neural tube with signals from the floor plate and notochord required for their differentiation into motoneurons.

The dependence of floor plate differentiation on inductive signals from the notochord makes it difficult to distinguish whether the D–V patterning of the neural tube is controlled by signals from the notochord, the floor plate, or both cell groups (Fig. 3). Signals from either the floor plate or the notochord may be sufficient to control the differentiation of distinct cell types in the ventral neural tube. This apparent redundancy may increase the fidelity with which the pattern of cell differentiation within the neural tube is established. At the neural plate stage, the notochord is in direct contact with the midline of the neural plate and may initiate the mediolateral patterning of the nervous system by inducing the floor plate. However, at subsequent stages of neurulation the notochord is displaced ventrally and becomes separated from the neural tube by mesenchymal cells (Jurand, 1962). Thus, if the notochord were the sole source of signals that control the pattern of cell differentiation along the D–V axis of the neural tube, the entire pattern would have to be established at a time when the notochord is still in proximity to the neural ectoderm. The acquisition of polarizing properties by the floor plate provides a mechanism by which midline-derived signals could continue to impose pattern on the

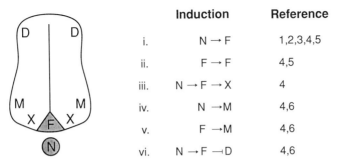

	Induction	Reference
i.	N → F	1,2,3,4,5
ii.	F → F	4,5
iii.	N → F → X	4
iv.	N → M	4,6
v.	F → M	4,6
vi.	N → F ⊣ D	4,6

Fig. 3. Summary of inductive interactions that have been demonstrated experimentally in the vertebrate spinal cord and hindbrain. N, notochord; F, floor plate; X, cells in X region; M, motor neurons; D, dorsal sensory relay neurons. (→) indicates induction, (⊣) indicates repression. Key to references: (1) van Straaten *et al.*, 1988; (2) Smith and Schoenwolf, 1989; (3) Placzek *et al.* 1990b. (4) Yamada *et al.*,1991; (5) Hatta *et al.*, 1991; (6) Placzek *et al.* 1990c.

adjacent neuroepithelium at stages after the displacement of the notochord.

INFLUENCE OF THE NOTOCHORD AND FLOOR PLATE ON CELL DIFFERENTIATION IN THE DORSAL NEURAL TUBE

The appearance of ventral cell types in dorsal regions after notochord grafts is accompanied by changes in the pattern of differentiation of dorsal cell types. Evidence for this is provided by analysis of the AC4 antigen which is restricted to the intermediate and dorsal regions of the spinal cord with the exception of a small region near to the roof plate by stage 17 (Yamada *et al.*, 1991). A marked decrease in expression of the AC4 antigen is observed in the dorsal spinal cord adjacent to notochord grafts (Yamada *et al.*, 1991). Similarly, the number of neurons that express a cellular retinoic acid binding protein (CRABP) (Maden *et al.*, 1989; Kitamoto *et al.*, 1989), a marker of spinal relay neurons, is decreased after lateral notochord grafts. Thus, after notochord grafts the induction of ventral cell types in dorsal regions is accompanied by an apparent suppression of the differentiation of dorsal cell types.

Analysis of the cell types remaining in the spinal cord after elimination of the notochord and floor plate has revealed a marked difference in the dependency of distinct neuronal classes on ventral midline-derived signals. Although motor neurons are completely absent, cell markers that are normally found in the dorsal spinal cord are present over an expanded D–V domain (Fig. 2d). The AC4 antigen which is normally absent from ventral region is expressed along the entire D–V extent of the spinal cord, except for the small region near the roof plate. In addition, neurons expressing CRABP are located at all positions including the ventral midline after notochord removal. The differentiation of the dorsal neural tube therefore does not appear to depend on the notochord and floor plate. The acquisition of dorsal cell properties could represent the constitutive fate of neural tissue cells. Alternatively, cells in the vental region of the neural tube may require floor plate and notochord-derived signals for their proliferation. In the absence of these signals, ventral cells may fail to divide, resulting in a deletion rather than a change in fate of cells in the ventral neural tube.

The differentiation of neuroepithelial cells into dorsal cell types may require signals from other cell groups, perhaps the roof plate. Indirect support for the existence of roof plate-derived signals that affect neural tube patterning comes from analysis of chick embryos which have received dorsal grafts of notochord or floor plate (Yamada et al., 1991). The induction of a floor plate in response to a dorsally placed notochord or floor plate occurs only when fusion of the dorsal midline of the neural tube fails, presumably preventing roof plate differentiation. Thus, signals from the roof plate may render the adjacent dorsal neural tube refractory to notochord and floor plate-derived signals. The nature of such roof plate-derived signals is not known but several members of the wnt family of growth factors are localized in and around the roof plate (Wilkinson et al., 1987; Roelink and Nusse, 1991).

POSSIBLE MECHANISMS FOR THE CONTROL OF NEURAL CELL PATTERN BY THE NOTOCHORD AND FLOOR PLATE

The ability of the floor plate and notochord to control the pattern of cell differentiation within the neural tube could be achieved in several different ways (Fig. 4).

First, contact-dependent signals might be initiated by the notochord and propagated by cells at the midline of the neural plate that later become the floor plate (Fig. 4i). During the time that it takes to propagate this signal laterally through the neural plate, the cells of the neural plate may gradually lose their competence to respond to this signal,

i   propagation of contact-dependent signal

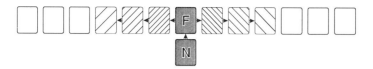

ii   diffusible signal acting on predisposed cells

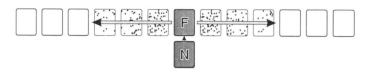

iii   gradient of diffusible signal

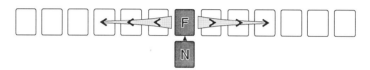

Fig. 4. Different mechanisms by which signals from the notochord (N) or floor plate (F) may generate and pattern distinct cell types in the ventral spinal cord. The diagram shows interactions occurring in the neural plate since it is likely that signals are initiated at this stage. (i) In this model, a contact-dependent signal initiated by the notochord is propagated by the floor plate. The lateral spread of this signal may be restricted by the gradual loss of competence of cells within the neural plate. (ii) In this model, signals operating at earlier stages of development have generated a predisposition in cells of the neural plate (designated by different intensities of stippling). A diffusible but nongraded signal from the floor plate or notochord may then act on more lateral cells in the neural plate to permit their differentiation into distinct cell types according to their predisposed fates. (iii) In this model a diffusible signal released by the floor plate or notochord establishes a concentration gradient that directly instructs cells as to their fate according to the concentration of signal to which they are exposed. For details see text.

thus resulting in different levels of the signal or graded responses to the signal which could lead to the formation of distinct cell types. Second, a diffusible signal which originates from the notochord floor plate may act on neural plate cells that are predisposed to different fates as a consequence of earlier signals (Fig. 4ii). In this case, the notochord and floor plate-derived signals would permit the differentiation of adjacent neural cells. Third, the notochord and floor plate could act as local sources of a factor that diffuses laterally through the neural plate, establishing a concentration gradient with its high point at the midline (Fig. 4iii). In this scheme, cell identity and the overall pattern of cell types would be defined by the concentration of diffusible signal to which precursor neuroepithelial cells are exposed (Yamada *et al.*, 1991). Neural plate cells exposed to highest concentrations of this factor would differentiate into floor plate; cells exposed to a lower concentration would differentiate into cells of region X and cells exposed to a still lower concentration would differentiate into motor neurons.

This third model is similar in principle to the way in which graded signals have suggested to generate cell pattern along the A–P axis of the developing chick wing bud (Tickle *et al.*, 1975). In the wing bud, A–P pattern appears to be under the control of a specialized region of posterior mesoderm known as the zone of polarizing activity (ZPA), which can evoke mirror-image digit duplications when grafted to ectopic sites (Tickle *et al.*, 1975). Retinoic acid mimics the effects of the ZPA (Tickle *et al.*, 1982) and appears to be distributed unevenly along the A–P axis of the limb bud with its highest concentration in the posterior mesenchyme (Thaller and Eichele, 1987). On this basis, it has been suggested that retinoic acid functions as an endogenous morphogen involved in establishing axial polarity in the developing chick limb (see Brockes, 1989; Eichele, 1989).

The notochord and floor plate, but not other regions of the neural tube, mimic the action of the ZPA and retinoic acid in respecifying digit pattern in the chick limb (Wagner *et al.*, 1990). Biochemical studies show that the floor plate can synthesize and release active retinoids *in vitro* (Wagner *et al.*, 1990, 1992), although at levels only marginally greater than those from the dorsal spinal cord. In addition, grafts of the ZPA or of retinoic acid-impregnated beads do not mimic the floor plate or notochord in changing the pattern of cell differentiation in the neural tube (T. Yamada, unpublished observations). Thus, at present, there is no evidence that the ability of the notochord and floor plate to control the pattern of cell differentiation in the neural tube involves retinoids, although it remains possible that retinoids released by the floor plate act in combination with additional factors.

Although these studies have focused on the control of D–V cell pattern, the notochord and floor plate may also contribute to neural cell differentiation along the A–P axis of the neural tube. Many classes of neurons are absent from the forebrain and restricted to more posterior regions of the central nervous system (CNS). For example, the most anterior group of motoneurons, which comprises the oculomotor nucleus, is located in the midbrain near the anterior end of the floor plate. Motoneurons represent one of several classes of neurons found at more posterior regions of the CNS which appear to be dependent on the floor plate for their differentiation. These observations reinforce the earlier suggestions of Ahlborn (1883) and Kingsbury (1930) that the A–P domain of the neuraxis over which motor neurons differentiate is defined by the notochord and floor plate.

Taken together, the results discussed in this chapter provide evidence that many aspects of cell patterning within the neural tube are controlled by signals that derive from two axial midline cell groups, the notochord and the floor plate. An early step in the establishment of cell pattern within the neural tube is the induction of the floor plate by the notochord. The floor plate may then regulate the pattern of cell differentiation along the D–V axis of the neural tube. Although the mechanisms underlying these patterning events have not been resolved, the identity of cells in the ventral region of the neural tube appears to be defined by their position with respect to the ventral midline. These observations support the idea that cell fate within the developing nervous system, as in other embryonic tissues and organisms (Ferguson and Anderson, 1992; Spemann, 1938; Gerhart and Keller, 1986; Tickle *et al.*, 1975), depends in large part on the interpretation of signals that derive from local sources in the embryo (Wolpert, 1969).

ACKNOWLEDGMENTS

We thank M. Placzek, M. Tessier-Lavigne, and T. Yamada for permission to cite their results. Vicki A. Leon and Ira Schieren helped in preparing the manuscript. T. M. Jessell is an Investigator and T. Yamada is a Research Associate of the Howard Hughes Medical Institute. J. Dodd was supported by grants from NIH and NSF, a Klingenstein Fellowship, and the Irma T. Hirschl Foundation.

REFERENCES

Ahlborn, F. (1883). Untersuchungen uber das Gehirn der Petromyzonten. Z. *Wiss. Zool.* **39**, 191–294.

Baker, R. C. (1927). The early development of the ventral part of the neural plate of amblystoma. *J. Comp. Neurol.* **44**, 1–27.

Bovolenta, P., and Dodd, J. (1990). Guidance of commissural growth cones at the floor plate in the embryonic rat spinal cord. *Development* **109**, 435–447.

Bovolenta, P., and Dodd, J. (1991). Perturbation of neuronal differentiation and axon guidance in the spinal cord of mouse embryos lacking a floor plate: Analysis of Danforth's short-tail mutation. *Development* **113**, 625–639.

Brockes, J. P. (1989). Retinoids, homeobox genes, and limb morphogenesis. *Neuron* **2**, 1285–1294.

Clarke, J. D. W., Holder, N., Soff, S. R., and Storm-Mathissen, J. (1991). Neuroanatomical and functional analysis of neural tube formation in notochordless Xenopus embryos: Laterality of the ventral spinal cord is lost. *Development* **112**, 499–516.

Dixon, J. E., and Kintner, C. R. (1989). Cellular contacts required for neural induction in Xenopus embryos: Evidence for two signals. *Development* **106**, 749–757.

Dodd, J., and Jessell, T. M. (1988). Axon guidance and the patterning of neuronal projections in vertebrates. *Science* **242**, 692–699.

Eichele, G. (1989). Retinoids and vertebrate limb pattern formation. *Trends Genet.* **5**, 246–251.

Eisen, J. S. (1991). Determination of primary motoneuron identity in developing zebrafish embryos. *Science* **252**, 569–572.

Ferguson, E. L., and Anderson, K. V. (1992). Decapentaplegic acts as a morphogen to organize dorsal–ventral pattern in the Drosophila embryo. *Cell* **71**, 451–461.

Fontaine-Perus, J., Chanconie, M., LeDouarin, N. M., Gershon, M. D., and Rothman, T. P. (1989). Mitogenic effect of muscle on the neuroepithelium of the developing spinal cord. *Development* **107**, 413–422.

Fraser, S., Keynes, R., and Lumsden, A. (1990). Segmentation in the chick embryo hindbrain is defined by cell lineage restrictions. *Nature* **344**, 431–435.

Gerhart, J., and Keller, R. E. (1986). Region-specific cell activities in amphibian gastrulation. *Annu. Rev. Cell Biol.* **2**, 201–229.

Grabowski, C. T. (1956). The effects of the excision of Hensen's Node on the early development of the chick embryo. *J. Exp. Zool.* **133**, 301–343.

Hamburger, V., and Hamilton, H. (1951). A series of normal stages in the development of chick embryo. *J. Morphol.* **88**, 49–92.

Hara, K. (1978). Spemann's organizer, in birds. *In* "Organizer—A Milestone of a Half-Century from Spemann." (O. Nakomura and S. Toivonen, eds.), pp. 221–265. Elsevier/North-Holland Biomedical Press, Amsterdam.

Hatta, K., Ho, R. K., Walker, C., and Kimmel, C. B. (1990). A mutation that deletes the floor plate and disturbs axonal pathfinding in Zebrafish. *Soc. Neurosci. Abstr.* **16**, 139.3.

Hatta, K., Kimmel, C. B., Ho, R. K., and Walker, C. (1991). The cyclops mutation blocks specification of the floor plate of the zebrafish central nervous system. *Nature* **350**, 339–341.

Hollyday, M., and Hamburger, V. (1977). An autoradiographic study of the formation of the lateral motor column in the chick embryo. *Brain Res.* **132**, 197–208.

Homma, S., Yaginuma, H., and Oppenheim, R. W. (1990). Cell death during the earliest stages of spinal cord development in the chick embryo. *Soc. Neurosci. Abstr.* **16**, 836.

Jacobson, A. G. (1981). Morphogenesis of the neural plate and tube. *In* "Morphogenesis and Pattern Formation" (T. G. Connelly *et al.*, eds.) pp. 233–263. Raven Press, New York.

Jacobson, A. G., Oster, G. F., Odel, G. M., and Chang, L. Y. (1986). Neurulation and the cortical tractor model for epithelial folding. *J. Embryol. Exp. Morphol.* **96**, 19–49.

Jacobson, C. O. (1964). Motor nuclei, cranial nerve roots and fiber pattern in the medulla oblongata after reversal experiments on the neural plate of axolotl larvae. I. Bilateral operations. *Zool. Bid. Uppsala* **36**, 73–160.

Jessell, T. M., Bovolenta, P., Placzek, M., Tessier-Lavigne, M., and Dodd, J. (1989). Polarity and patterning in the neural tube: The origin and role of the floor plate. *Ciba Found. Symp.* **144**, 255–280.

Jurand, A. (1962). The development of the notochord in chick embryos. *J. Embryol. Exp. Morphol.* **10**, 602–621.

Kimmel, C. B., Hatta, K., and Eisen, J. S. (1991). Control of primary neuronal development in zebrafish. *Development Suppl. Genetic* **2**, 47–57.

Kingsbury, B. F. (1930). The developmental significance of the floor plate of the brain and spinal cord. *J. Comp. Neurol.* **50**, 177–207.

Kitamoto, T., Momoi, M., amd Momoi, T. (1989). Expression of cellular retinoic acid binding protein II (chick-CRABP II) in the chick embryo. *Biochem. Biophys. Res. Commun.* **164**, 531–536.

Kuwada, J. Y., and Hatta, K. (1990). Axonal pathfinding in the spinal cord of zebrafish mutants missing the floor plate. *Soc. Neurosci., Abstr.* **16**, 139.1.

Kuwada, J. Y., Bernhardt, R. R., and Chitnis, A. B. (1990a). Pathfinding by identified growth cones in the spinal cord of zebrafish embryos. *J. Neurosci.* **10**, 1299–1308.

Kuwada, J. Y., Bernhardt, R. R., and Nguyen, N. (1990b). Development of spinal neurons and tracts in the zebrafish embryo. *J. Comp. Neurol.* **302**, 617–628.

Langman, J., and Haden, C. C. (1970). Formation and migration of neuroblasts in the spinal cord of the chick embryo. *J. Comp. Neurol.* **138**, 419–432.

Leber, S. M., Breedlove, S. M., and Sanes, J. R. (1990). Lineage, arrangement, and death of clonally related motoneurons in chick spinal cord. *J. Neurosci.* **10**, 2451–2462.

Maden, M., Ong, D. E., Summerbell, D., Chytil, F., and Hirst, E. A. (1989). Cellular retinoic acid-binding protein and the role of retinoic acid in the development of the chick embryo. *Dev. Biol.* **135**, 124–132.

Mangold, O. (1933). Uber die Induktionsfahigkeit der verschiedenen Bezirk der Neurula von Urodelen. *Naturwissenschaften* **21**, 761–766.

Martins-Green, M. (1988). Origins of the dorsal surface of the neural tube by progressive delamination of epidermal ectoderm and neuroepithelium: Implications for neurulation and neural tube defects. *Development* **103**, 687–706.

McKanna, J. A., and Cohen, S. (1989). The EGF receptor kinase substrate p35 in the floor plate of embryonic rat CNS. *Science* **243**, 1477–1479.

Placzek, M., Tessier-Lavigne, M., Jessell, T. M., and Dodd, J. (1990a). Orientation of commissural axons in vitro in response to a floor plate derived chemoattractant. *Development* **110**, 19–30.

Placzek, M., Tessier-Lavigne, M., Yamada, T., Dodd, J., and Jessell, T. M. (1990b). The guidance of developing axons by diffusible chemoattractants. *Cold Spring Harbor Symp.* **55**, 279–289.

Placzek, M., Tessier-Lavigne, M., Yamada, T., Jessell, T. M., and Dodd, J. (1990c). Mesodermal control of neural cell identity: Floor plate induction by the notochord. *Science* **250**, 985–988.

Puelles, L., Ariat, J. A., and Martinez-de-la-Torre, M. (1987). Segment related mosaic neurogenetic pattern in the forebrain and mesencephalon of early chick embryos. 1. Topography of AChE-positive neuroblasts up to stage HH18. *J. Comp. Neurol.* **266**, 247–268.

Roach, F. C. (1945). Differentiation of the central nervous system after axial reversals of the medullary plate of amblystoma. *J. Exp. Zool.* **99**, 53–77.

Roelink, H., and Nusse, R. (1991). Expression of two members of the wnt family during mouse development—restricted temporal and spatial patterns in the developing neural tube. *Genes Dev.* **5**, 381–388.

Rosenquist, G. C. (1966). A radioautographic study of labelled grafts in the chick blastoderm. *Contrib. Embryol. Carnegie Inst.* **38**, 71–110.

Rosenquist, G. C. (1983). The chorda center in Hensen's node of the chick embryo. *Anat. Rec.* **207**, 349–355.

Ruiz i Altaba, A. (1990). Neural expression of the Xenopus homeobox gene Xhox3: Evidence for a patterning neural signal that spreads through the ectoderm. *Development* **108**, 595–604.

Ruberte, E., Dolle, P., Chambon, P., and Morriss-Kay, G. (1991). Retinoic acid receptors and cellular retinoid binding proteins. II. Their differential pattern of transcription during early morphogenesis in mouse embryos. *Development* **111**, 45–60.

Schoenwolf, G. C., and Smith, J. L. (1990). Mechanisms of neurulation: Traditional viewpoint and recent advances. *Development* **109**, 243–270.

Schoenwolf, G. C., Folsom, D., and Moe, A. (1988). A reexamination of the role of microfilaments in neurulation in the chick embryo. *Anat. Rec.* **220**, 87–102.

Selleck, M. A. J., and Stern, C. D. (1991). Fate mapping and cell lineage analysis of Hensen's node in the chick embryo. *Development* **112**, 615–626.

Smith, J. L., and Schoenwolf, G. C. (1989). Notochordal induction of cell wedging in the chick neural plate and its role in neural tube formation. *J. Exp. Zool.* **250**, 49–62.

Spemann, H. (1938). "Embryonic Development and Induction." Yale Univ. Press, New Haven.

Steding, G. V. (1962). Experiments zur morphogenese des Ruckenmarkes. *Acta Anat.* **49**, 199–231.

Tanaka, H., and Obata, K. (1984). Developmental changes in unique cell surface antigens of chick embryo spinal motor neurons and ganglion cells. *Dev. Biol.* **106**, 26–37.

Tessier-Lavigne, M., Placzek, M., Lumsden, A. G. S., Dodd, J., and Jessell, T. M. (1988). Chemotropic guidance of developing axons in the mammalian central nervous system. *Nature* **336**, 775–778.

Thaller, C., and Eichele, G. (1987). Identification and spatial distribution of retinoids in the developing chick limb bud. *Nature* **327**, 625–628.

Theiler, K. (1959). Anatomy and develoment of the "truncate" (boneless) mutation in the mouse. *Am. J. Anat.* **104**, 319–343.

Tickle, C., Alberts, B., Wolpert, L., and Lee, J. (1982). Local application of retinoic acid to the limb bud mimics the action of the polarizing region. *Nature* **296**, 564–566.

Tickle, C., Summerbell, D., and Wolpert, L. (1975). Positional signalling and specification of digits in chick limb morphogenesis. *Nature* **254**, 199–202.

van Straaten, H. W. M., and Drukker, J. (1987). Influence of the notochord on the morphogenesis of the neural tube. *In* "Mesenchymal–Epithelian Interactions in Neural Development" (J. R. Wolff *et al.*, eds.) pp. 153–162. NATO ASI Series Vol. H5 Springer-Verlag, Berlin.

van Straaten, H. W. M., Hekking, J. W. M., Beursgens, J. P. W. M., Terwindt-Rouwenhorst, E., and Drukker, J. (1989). Effect of the notochord on proliferation and differentiation in the neural tube of the chick embryo. *Development* **107**, 793–803.

van Straaten, H. W. M., Hekking, J. W. M., Thorn, F., Wiertz-Hoessels, E. L. M., and Drukker, J. (1985a). Induction of an additional floor plate in the neural tube. *Acta Morphol. Neerl. Scand.* **23**, 91–97.

van Straaten, H. W. M., Hekking, J. W. M., Wiertz-Hoessels, E. L., Thors, F., and Drukker, J. (1988). Effect of the notochord on the differentiation of a floor plate area in the neural tube of the chick embryo. *Anat. Embryol.* **177**, 317–324.

van Straaten, H. W. M., Thors, F., Wiertz-Hoessels, E. L., Hekking, J. W. M., and Drukker, J. (1985b). Effect of a notochordal implant on the early morphogenesis of the neural tube and neuroblasts: Histometrical and histological results. *Dev. Biol.* **110**, 247–254.

Wagner, M., Han, B., and Jessell, T. M. (1992). Regional differences in retinoid release from embryonic neural tissue detected by an *in vitro* reporter assay. *Development* **116**, 55–66.

Wagner, M., Thaller, C., Jessell, T., and Eichele, G. (1990). Polarizing activity and retinoid synthesis in the floor plate of the neural tube. *Nature* **345**, 819–822.

Wallace, J. A. (1985). An immunocytochemical study of the development of central serotonergic neurons in the chick embryo. *J. Comp. Neurol.* **236**, 444–453.

Watterson, R. L., Goodheart, C. R., and Lindberg, G. (1955). The influence of adjacent structures upon the shape of the neural tube and neural plate of chick embryos. *Anat. Rec.* **122**, 539–559.

Wilkinson, D. G., Bailes, J. A., and McMahon, A. P. (1987). Expression of the proto-oncogene int-1 is restricted to specific neural cells in the developing mouse embryo. *Cell* **50**, 79–88.

Wolff, E. (1936). Les bases de la teratogenese experimentale des vertebres amniotes d'apres les resultats de method, directes. *Arch. Anat. Histol. Embryol.* **22**, 1–375.

Wolpert, L. (1969). Positional information and the spatial pattern of cellular differentiation. *J. Theoret. Biol.* **25**, 1–47.

Yaginuma, H., Homma, S., Kunzi, R., and Oppenheim, R. W. (1991). Pathfinding by growth cones of commissural interneurons in the chick embryo spinal cord: A light and electron microscopic study. *J. Comp. Neurol.* **304**, 78–102.

Yaginuma, H., Shiga, T., Homma, S., Ishihara, R., and Oppenheim, R. W. (1990). Identification of early developing axon projections from spinal interneurons in the chick embryo with a neuron specific β-tubulin antibody: Evidence for a new "pioneer" pathway in the spinal cord. *Development* **108**, 705–716.

Yamada, T., Placzek, M., Tanaka, H., Dodd, J., and Jessell, T. M. (1991). Control of cell pattern in the developing nervous system: Polarizing activity of the floor plate and notochord. *Cell* **64**, 635–647.

11

Retinoids and Pattern Formation in Vertebrate Embryos

OLOF SUNDIN,[1] REINHOLD JANOCHA,[2] AND
GREGOR EICHELE
V. and M. McLean Department of Biochemistry
Baylor College of Medicine
Houston, Texas 77030

BIOCHEMISTRY OF RETINOIDS

The molecular and cellular processes that establish the pattern of vertebrates are currently under intensive investigation (for recent reviews, see 1–3). Broadly speaking, morphogenesis in a multicellular system depends on two type of signaling mechanisms: short-range interactions between neighboring cells (e.g., 4) and chemical signals that diffuse over distances of a few hundred microns, i.e., a dimension encompassing several cell diameters. Even a cursory glance at an embryo reveals an intricate tissue organization and it does not require a big leap of faith to envision direct interactions between different tissues. The challenge is to unravel such interactions at cellular and molecular levels and to demonstrate that they are causal to morphogenesis.

Whether long-range chemical signals exist has been debated since the inception of modern experimental embryology. The rapid progression of embryonic development and the laws of diffusion in a tissue environment limit the size of a diffusible signaling substance to a few

---

[1] *Present address:* Wilmer Eye Institute, The Johns Hopkins University School of Medicine, 600 N. Wolfe Street, Baltimore, MD 21205.

[2] *Present address:* Department of Toxicology and Pathology, Hoffmann-La Roche, Nutley, NJ 07110.

Cell–Cell Signaling in
Vertebrate Development

kilodaltons or less (5). Hence, in multicellular systems putative diffusible signaling molecules are probably not large proteins but smaller molecules either that are able to progress directly from cell to cell or for which cells have cell surface receptors. The small size of embryos and the presumably low concentrations of signaling substances are two major difficulties in the search for morphogenetic signaling molecules.

In adult organisms, however, numerous small signaling molecules play an important role in cell physiology. These agents include, for example, prostaglandins, leukotrienes, steroid hormones, and retinoids. It is reasonable to assume that at least some of these signaling molecules also operate during ontogeny. Steroids, retinoids, and other related hormones bind to specific nuclear receptors (6–8). These receptors form a multigene family and are found throughout the animal kingdom (9). They are multidomain proteins that consist, in essence, of a transactivation region, a DNA-binding domain, and a ligand-binding region. When hormone binds to the ligand-binding domain, the receptor is activated and functions as a transcription factor that regulates the expression of specific target genes. Target genes feature specific receptor-binding sites (so-called hormone-responsive elements) that have a high affinity for a particular receptor or set of receptors. Mechanistically, the sequence of events by which a hormone binds to its receptor and thereby activates gene transcription can be conceptualized in terms of a straightforward paradigm, the signal transduction process.

Hormone-dependent gene regulation can be extraordinarily complex. all-*trans*-Retinoic acid, for example, binds to three different receptors known as α-RAR (RAR; retinoic acid receptor), β-RAR, and γ-RAR (see also below for more information about retinoid receptors) (10–15). Each of these receptors is encoded by a separate gene. A second layer of complexity results from the fact that RAR genes produce several transcripts generated by alternative promoter usage and/or by alternative splicing (e.g., 16–18). In addition to the retinoic acid receptors there is also a second class of retinoid receptors, referred to as the RXRs (19–23). There are three RXRs (α-RXR, β-RXR, and γ-RXR). RXRs can bind to and be activated by 9-*cis*-retinoic acid (24, 25). Moreover, while RXRs do not bind all-*trans*-retinoic acid, RARs will bind both all-*trans*- and 9-*cis*-retinoic acid. A third layer of complexity is added by the fact that RAR and RXR subunits can interact to form heterodimers with altered regulatory functions (21, 23, 26, 27). The precise implications of heterodimers for transcriptional regulation are not yet clear, but they suggest interactions between the RAR and RXR signaling pathways.

Studies conducted over the past few years in several laboratories have established that retinoid receptors are broadly expressed in vertebrate embryos often in intriguing and complex patterns (22, 28–38). It is reasonable to assume that such diverse expression patterns of RARs and RXRs reflect an important function of retinoids in various fundamental aspects of embryogenesis such as positional signaling and cell differentiation. There is good evidence that retinoids, including all-*trans*-retinoic acid, are present in embryonic tissues (40–42). The most detailed studies on endogenous retinoids are those in chick limb buds, where so far a total of six retinoids have been identified (Fig. 1) (40, 42). As indicated by arrows in Fig. 1, endogenous retinoids are interconverted *in vivo* by specific enzymatic reactions (43, 44; and Janocha and Eichele, in preparation). For example, retinol is oxidized to retinal by dehydrogenases found in microsomal preparations of chick embryonic tissue (Fig. 2). Retinal is further oxidized to retinoic acid by a second dehydrogenase located in cytosol (reviewed in 44). The degradation of retinoic acid begins with its metabolism to 4-oxoretinoic acid. This reaction occur in microsomes and can be inhibited by carbon monoxide, metyrapone, and SKF-525A, suggesting that the reaction is mediated by a cytochrome P-450 enzyme (45). 3,4-Didehydroretinoic acid is a second morphogenetically active retinoid which is also found endogenously in chick limb buds (42). Recent work has shown that in embryonic chick tissues 3,4-didehydroretinoids are primeraly generated from retinol (Fig. 1 and Ref. 42), but the nature of the enzyme(s) responsible for the introduction of a double bond at the 3,4 position is elusive. It will be necessary to purify the retinoid metabolism enzymes to homogeneity and with the help of *in situ* hybridization and specific antibodies characterize their spatial distribution within the embryo. An attractive possibility is that synthesis

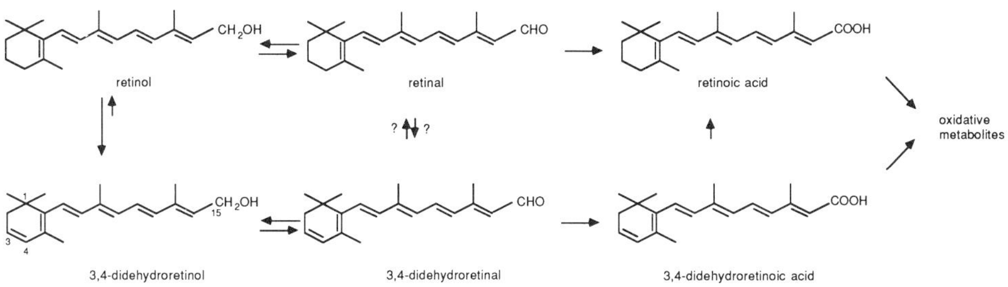

Fig. 1. Reaction pathways leading from all-*trans*-retinol to all-*trans*-retinoic acid and 3,4-didehydro-all-*trans*-retinoic acid, respectively. Note, all-*trans*-retinoic acid and its didehydro analog are metabolized to various oxidative metabolites (45). How the RXR ligand 9-*cis*-retinoic acid relates to the scheme shown is not yet known.

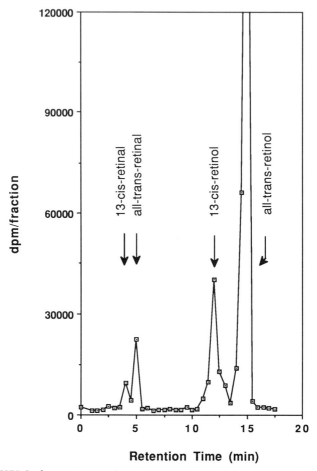

Fig. 2. HPLC chromatogram showing conversion of all-*trans*-retinol to all-*trans*-retinal by microsomal fractions isolated from 3.5-day-old chicken embryos (stage 20/21). Fifty micrograms of microsomal protein prepared by differential centrifugation was incubated with all-*trans*-retinol (500 nM, pH 7.4, 2 mM NAD+). The minor peaks seen at 3.8 and 11.8 min, respectively, are 13-*cis*-retinal and 13-*cis*-retinol that had presumably formed during the extraction procedure.

and degradation of retinoic acid are spatially segregated. This would result in a spatial gradient of retinoic acid of the kind found in the developing limb bud (40). There is some evidence for spatial segregation of retinoid metabolism in the neural tube (46). The activity that converts retinol to 3,4-didehydroretinol is enriched in tissue of the floor plate, a distinct structure which occupies the ventral midline of the neural tube (47). This enrichment is by a factor of about 3 to 4 when compared to the activity in tissue of the lateral or dorsal neural tube (46). This observation is particularly interesting in view of the finding that floor plate but not other regions of the neural tube is capable of acting as polarizing tissue in the limb bud system (46).

Other components of the retinoid signaling pathway are the cellular retinoid-binding proteins that have high affinity for retinol or retinoic acid (reviewed in 48). There are two cellular retinol-binding proteins (CRBP I and II) encoded by distinct genes and two cellular retinoic acid-binding proteins (CRABP I and II), also encoded by different genes. Several investigators have observed regionalized expression of these retinoid-binding proteins in embryonic tissues. A good illustration of highly localized CRABP expression comes from studies in the developing hindbrain in mouse and chicken embryos (32, 49–51). For example, Maden *et al.* (52) demonstrated that CRABP I protein is already present in the neural tube of early chicken embryos immediately after neural fold closure, as well as in the neural crest and its derivatives. Later in development, at a period when the segmental rhombomeres are established, CRABP I-positive cells appear in three distinct locations of the central nervous system. First, in the developing spinal cord through the posterior rhombencephalon up to rhombomere 6, second in rhombomere 4, and third in the roof of the anterior mesencephalon. The gaps of CRABP I immunoreactivity are gradually filled in, so that at later stages the labeled neuroblasts extend continuously from the mesencephalon all the way to the caudal end of the neural tube. The spatial and temporal change of CRABP expression in the rhombomeres is reminiscent of the neurofilament staining patterns which reflect differences in the timetable of neurogenesis between alternate rhombomeres (53) and of the rhombomere-restricted expression patterns of the *Hox-2* genes (e.g., 54 and below). It is interesting to note that the distribution of CRABP overlaps that of *Hox-2.9* in the hindbrain (52). Such expression patterns are perhaps clues to the site of action of retinoid-binding proteins, but do not reveal their biochemical function. One plausible function is that binding proteins operate as buffers that regulate the concentration of free ligand. Experimental evidence for this view comes, for example, from studies in which CRAPB I was overexpressed in cultured F9 teratocarcinoma cells. F9 cells differentiate when exposed to retinoic acid (55). It was found that F9 cells overexpressing CRABP I require more retinoic acid to differentiate than F9 cells with normal levels of CRABP I (56). Another likely function of binding proteins is participation in the regulation of retinoid metabolism. Ong and co-workers (57) have discovered that the rate of esterification of retinol to retinylesters is substantially enhanced by the presence of CRBP II.

The main purpose of retinoids, binding proteins, and receptors is to regulate the expression of target genes. Hence the identification of such target genes is central to understanding retinoid physiology. Chytil and ul-Haq have recently compiled a list of more than 100 genes and gene products whose expression is affected by retinoic acid (58). However, it is as yet unclear what fraction of these genes responds to retinoic acid directly. As pointed out above, the cellular responses to retinoic acid are mediated by two families of transcription factors, RARs and RXRs. Retinoid receptors are ligand-dependent transcription factors that bind to specific, conserved DNA sequences. These binding sequences, termed hormone response elements are composed of direct repeat, palindromic, and inverted palindromic arrangements of the core motif "AGGTCE," thereby imparting selective responses to different receptors. Receptor specificity is further determined by the "spacer" of nucleotides between each motif (21, 23, 59–61). To give an example, spacers of 3, 4, and 5 nucleotides between direct repeats of the core sequence confer response to vitamin D, thyroid hormone, and retinoic acid, respectively. Thus a spacing rule provides a way to selectively address distinctive regulatory networks. Recently, a RXR specific responsive element was found in the promoter of the CRBP II. This element consists of five nearly perfect tandem repeats of the AGGTCA core element (61). Interestingly, the RXR-dependent activation of the CRBP II gene is down-regulated in the presence of RAR. The reason for this is that RAR is capable of binding tightly to the CRBP II—RXRE and thereby blocks CRBP II activation by RXR (61). RXR-dependent CRBP II activation also leads to the notion that retinoic acid might regulate its own metabolism (61). Additional evidence for this view derives from reports that CRABP II and the alcohol dehydrogenase ADH3 gene (a gene encoding a protein that might convert retinol to retinal) are also retinoic acid responsive (62–64).

The homeobox genes of the Antennapedia-Bithorax family constitute a developmentally important family of genes whose expression in vertebrates is influenced by retinoic acid. In both insects and vertebrates this highly conserved family of transcription factors plays a key role in determining the position of structures along the developing body axis (reviewed in Ref. 65). The first evidence that expression of these genes can be influenced by retinoic acid came from studies in which mouse teratocarcinoma cells were induced to differentiate by treatment with retinoic acid. Levels of homeobox mRNAs were found to rise at least 10-fold during the course of this differentiation process (66–69). While a number of homeobox genes respond to treatment with retinoic acid, it has been found that the rate of response varies considerably and is related to the genomic organization of these genes. The vertebrate Antennapedia-Bithorax homeobox genes are arranged in four clusters located on different chromosomes (65). Between clusters the gene order is highly conserved and transcription units are all oriented in the same direction. The remarkable rule which has emerged from these studies is that the genes at the 3' end of each cluster, all of which share close homology with the *Drosophila* gene *labial*, are the first to be activated by retinoic acid treatment (69–72). La Rosa and Gudas were the first to observe that induction of a homeobox gene of the *labial* type, *Hox-1.6*, could occur within a few hours after treatment, well prior to the onset of retinoid-induced differentiation of the teratocarcinoma cells (70). Over a period of several days the other homeobox genes are activated sequentially as one proceeds from the 3' to the 5' end of the locus (69, 71, 72). Another feature of this sequence is that in general it reflects the order in which the genes are activated within the developing embryo (71, 72).

The mechanism for the induction of homeobox genes by retinoic acid is not yet well understood, and opposing views have been advanced for post-transcriptional control (66), or for regulation at the level of transcription (70, 72). Concerning the issue of whether these genes are indirectly or directly regulated by retinoids, there is evidence that the induction of *Hox-1.6* (70), as well as other *Hox* genes (72), is independent of protein synthesis. This is consistent with direct activation of the genes through retinoid receptors, but this issue cannot be fully resolved until the retinoic acid-responsive elements in the genome have been identified. Interestingly, inhibition of protein synthesis allows retinoic acid to rapidly and simultaneously induce all genes in the locus within a few hours, including those toward the 5' end (72). Normally, the 5' genes require several days to reach maximal levels of induction.

RETINOIDS INDUCE A HOMEOBOX GENE DURING EARLY DEVELOPMENT OF THE CHICK EMBRYO

The observations made in cultured teratocarcinoma cells raise the question of whether a similar phenomenon also takes place in the embryo. Several investigators have examined the expression pattern of homeobox genes in embryos treated with retinoic acid (e.g., 73–79).

The chick embryo is well suited to address such questions since it can be removed from the egg and cultured in the presence of retinoic acid or other reagents (a detailed account of these findings can be found in Ref. 74). Retinoic acid treatment of primitive streak chick embryos shows few immediate effects, but results in pronounced morphological changes as development proceeds to later stages. Embryos are smaller, there is less mesodermal tissue, they lack a properly developed heart (39), and there are considerable changes in the organization of the central nervous system (Fig. 3). The anteriormost structures such as forebrain and eye primordia are somewhat diminished in size, but the most striking defect is the absence of a normal midbrain and anterior hindbrain. In the place of these structures there is only a short portion of neural tube (region between dashed lines in Fig. 3D). At this stage, eight segmental units known as rhombomeres are normally visible as a series of constrictions along the axis of the hindbrain (Fig. 3B). These constrictions are not well defined in the retinoic acid-treated embryos, and assignments of position within this region must be made on the basis of other criteria. An especially useful marker is the massive cluster of neural crest cells (labeled with "g" in Fig. 3D) attached to the anteriormost section of the hindbrain-like neuroectoderm (Figs. 3C and 3D). It has the characteristic morphology of the 7th/8th ganglion primordium which is derived from and attached to rhombomere 4 (Figs. 3A and 3B). This portion of neuroectoderm also expresses the rhombomere 4-specific Ghox 2.9 marker (80, see below). One interpretation of these results is that the entire midbrain and anterior half of the hindbrain (rhombomeres 1 through 3) have been transformed into the small segment situated between the forebrain and rhombomere 4. In contrast to this, it appears that rhombomere 4 and the more posterior portions of the hindbrain have been spared and are even enlarged.

The most sensitive period for this dysmorphogenesis is prior to and during late gastrulation (Hamburger–Hamilton stage 4). Treatments beginning a few hours later, during the neurula stage (stage 6), are no longer effective in causing major defects in the central nervous system. Overt morphological changes in the central nervous system, however, are not evident until much later, when the neural tube becomes partitioned into segments (stages 9 through 11). This raises the question of what changes occur in the embryos at earlier stages and how these produce the malformations observed later on. In an effort to examine this issue at molecular and cellular levels, we have investigated how Ghox 2.9 expression is affected by retinoic acid treatment.

When stage 4 embryos are treated for 4 hr with 6 μM retinoic acid they continue to develop in a relatively normal manner to early stage 5.

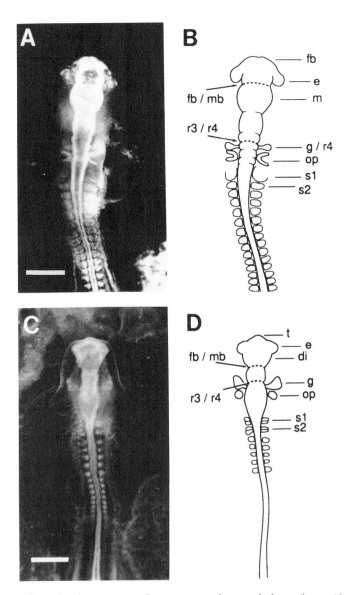

Fig. 3. Effect of early retinoic acid treatment on the morphology of stage 12 embryos. (A) Control embryo grown in culture from stage 4 to stage 12, fixed, photographed in dimethyl sulfoxide (DMSO)/methanol (1 : 1) with incident illumination. Dorsal view. (B) Interpretive drawing of A. Dashed lines indicate the forebrain/midbrain boundary and the anterior boundary of rhombomere 4, respectively. (C) Experimental embryo, treated with 6 μM retinoic acid for 4 hr. (D) Interpretive drawing of C. Dashed lines are as in B. di, diencephalon; e, eye; fb, forebrain; fb/mb, forebrain/midbrain boundary; g ≠ r4 or g (in D), primordium of acousticofacial ganglion derived from rhombomere 4 neural crest; m, midbrain; op, otic placode; r3/r4, anterior boundary of rhombomere 4; s1, the somite which first appears at stage 7; s2, second somite; t, telencephalon. Bars: 400 μm.

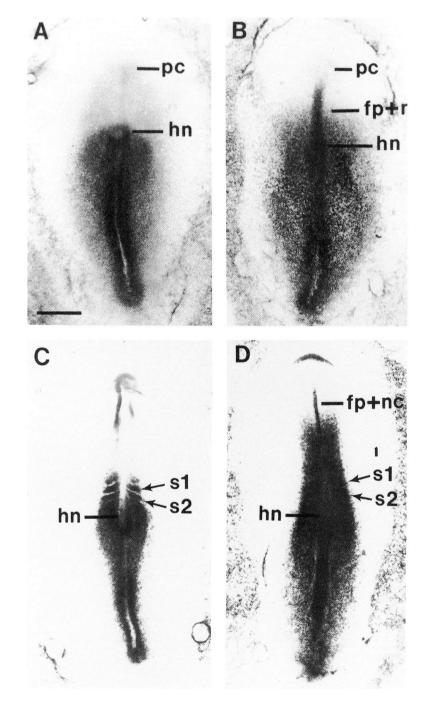

However, when these embryos are stained to reveal the expression of Ghox 2.9 protein, one observes number of changes (compare Figs. 4A and 4B). There is significant ectopic expression of Ghox 2.9 protein in anterior regions (Fig. 4B) where the gene is normally not active (Fig. 4A). Hensen's node and midline structures extending up to the pre-chordal plate show a striking increase in Ghox 2.9 antibody labeling (Fig. 4B), while there appears to be no induction in the prechordal plate or its overlying ectoderm. Ghox 2.9 is also induced in the region

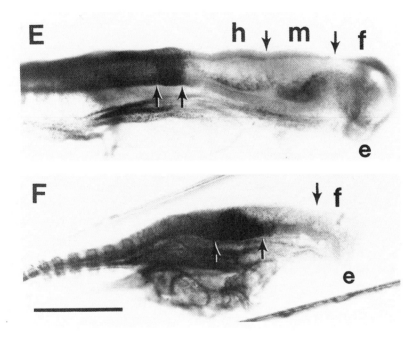

Fig. 4. Effect of retinoic acid on Ghox 2.9 expression. Embryos grown in cluture were fixed at different times after retinoic acid treatment and whole-mount stained for Ghox 2.9 protein. Embryos were fully cleared in benzyl alcohol/benzyl benzoate. (A) Control embryo, stage 5. (B) Retinoic acid-treated embryo, stage 5. (C) Control embryo, late stage 7, with two somites and neural plate. (D) Retinoic acid-treated, late stage 7, with two somites and neural plate. (E) Control embryo, late stage 11, lateral aspect, anterior to right. (F) Retinoic acid-treated embryo, stage 11, lateral aspect, ante-rior to right. In E and F, arrows pointing up indicate anteroposterior extent of rhomb-omere 4-specific Ghox 2.9 signal. In E the boundaries between hind-, mid- and forebrain are marked by downward arrows. In F, the posterior forebrain boundary is indicated by an arrow pointing down. e, eye; hn, Hensen's node; fp+nc, floor plate plus notochord; pc, prechordal plate; s1, first somite; s2 second somite. Bar in 4A: 400 μm (A–D). Bar in 4F: 400 μm (E and F).

lateral to the midline up to a level halfway between Hensen's node and the prechordal plate. Transverse sections of treated embryos reveal that the ectopic induction of Ghox 2.9 takes place primarily in the ectoderm, which is committed to become the midline floor plate and lateral neural plate (see Figs. 5A–5D). The section situated just posterior to the prechordal plate (Fig. 5A) shows the selective induction of Ghox 2.9 in the floor plate, with less in the underlying notochord. At a more posterior level, just anterior to Hensen's node, both floor plate and presumptive lateral neuroectoderm have been induced to express high levels of Ghox 2.9 (Figs. 5B). Underlying mesodermal structures, such as notochord and lateral plate mesoderm, show some response to retinoic acid, but the degree of Ghox 2.9 induction is substantially lower than that in the overlying ectoderm. The section through Hensen's node (Fig. 5C) reveals that at this more posterior level all tissues, including the mesoderm, exhibit considerable induction of Ghox 2.9 protein relative to an untreated embryo (Fig. 5D).

When retinoic acid-treated embryos are withdrawn from retinoic acid and allowed to reach the two-somite stage (stage 7), they continue to develop with a normal timetable of notochord elongation, node regression, and somite formation. The ectoderm and mesoderm of treated embryos are slightly thinner, but key landmarks apper in their normal location and display normal spacing relative to each other. Nevertheless, the pattern of Ghox 2.9 expression remains markedly different from that of untreated embryos. In the normal embryo, Ghox 2.9 expression terminates a short distance anterior to the first somite (Fig. 4C). In contrast, Fig. 4D makes it very clear that in a treated embryo the Ghox 2.9 domain extends far anterior to somite 1. In this ectopic domain, Ghox 2.9 signal is concentrated almost entirely within the neuroectoderm, with only modest induction within the mesoderm (Fig. 5F). Ectopic expression is also observed along the midline almost up to the level of the prechordal plate (Fig. 4D). Examination of cross sections reveals that this is due to induction of Ghox 2.9 in both the floor plate and the underlying notochord (Fig. 5E). More posteriorly, at the level of the first somite, retinoic acid treatment results in considerable expression in floor plate and notochord (Fig. 5G), tissues which are largely Ghox 2.9-negative at this position in the control embryo (Fig. 5H). At stage 7, treated embryos still show enhanced amounts of Ghox 2.9 protein within its normal domain of expression (Figs. 5G and 6H, see neural plate and mesoderm).

When these embryos are grown even further, to stage 11, they exhibit malformations of the type illustrated in Figs. 3C and 3D. In these embryos, the enlarged hindbrain-like neuroepithelium is positive for

Ghox 2.9 (Fig. 4F) and in its anterior portion there is a band of intense Ghox 2.9 expression (Fig. 4F, region between the two arrows pointing up) which corresponds to an enlarged rhombomere 4. In this embryo rhombomere 4 is located at a relatively more anterior position along the body axis and appears to occupy a larger than normal portion of the neural tube. These data raise the possibility that the early ectopic expression of Ghox 2.9 reflects an early transformation of cell fate which underlies the abnormalities which appear later in the nervous system.

RETINOIC ACID INDUCES SOMITE-LIKE STRUCTURES IN CRANIAL MESODERM

In addition to its effects on the central nervous system, retinoic acid causes unusual alterations in the development of the lateral plate mesoderm. During normal embryogenesis the first somite is cleaved from the continuous mesoderm lateral to the notochord and successive somites are formed at the rate of one per hour in a rostral to caudal fashion. In embryos treated for 4 hr in 6 μM retinoic acid at stage 4, the first two somites appear at the normal time and location (Fig. 4D). Later, in the three-somite embryo, several small somite-like structures simultaneously appear in the lateral plate mesoderm anterior to the native somites (Figs. 6B and 6C). In normal embryos the mesoderm of this region shows no obvious segmentation (Fig. 6A). While these structures are small, are variable in number, and appear independently of the normal rostral-to-caudal somite sequence, sections reveal an epithelial cell arrangement which is highly reminiscent of true somites (not shown). In retinoic acid-treated mouse embryos it has been observed that the somite column extends to a more anterior position than normal (79). It is not clear, however, whether these mouse somites arise in a normal rostrocaudal sequence or on a separate timetable.

One intriguing possibility is that these anterior structures are related to the somitomeres, subtle features which are thought to reveal a segmental organization of the lateral plate mesoderm in the head (81). It has also been suggested that somites of the trunk are first organized in the lateral plate mesoderm as somitomeres and become readily visible as somites when these undergo a transformation to epithelial mesoderm. The key feature in which anterior somitomeres differ from trunk somitomeres is that they do not undergo this epithelial transformation. One might speculate that retinoic acid somehow releases a latent tendency of the anterior somitomeres to condense as epithelial

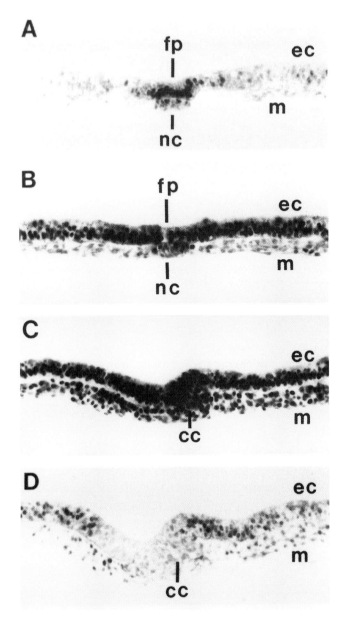

Fig. 5. Tissue-specific induction of Ghox 2.9 by retinoic acid. Embryos corresponding to those in Figs. 4A–4D were sectioned transversely, and sections were photographed in bright-field. Dorsal is up. (A–C) Retinoic acid treated, fixed at stage 5. (A) Anterior section, near prechordal plate. (B) Section anterior to Hensen's node. (C) Section through Hensen's node. Chorda center is asymmetric, mesodermal component shifted to the right. (D) Control embryo cultured without retinoic acid, section passes through the node. Chorda center mesoderm shifted to the right. (E–G) Retinoic

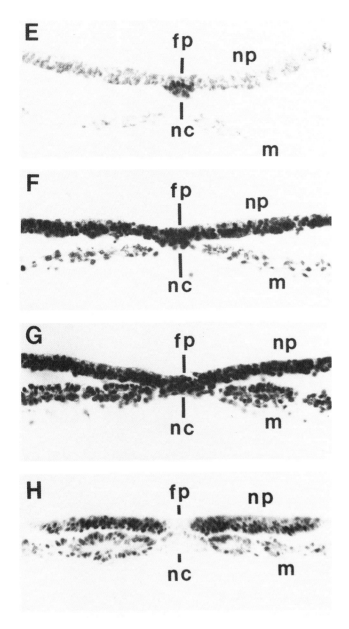

acid treated, fixed at late stage 7. (E) Anterior section, near prechordal plate. Note: stain in the lateral neural plate is largely background, with some weakly positive nuclei. (F) Section midway between first somite and prechordal plate. (G) Section at level of first somite. (H) Control embryo, cultured without retinoic acid, fixed at late stage 7, and sectioned at level of first somite. cc, Chorda center; ec, ectoderm; fp, floor plate; m, lateral plate mesoderm; nc, notochord; np, neural plate. Bar: 50 μm (for entire figure).

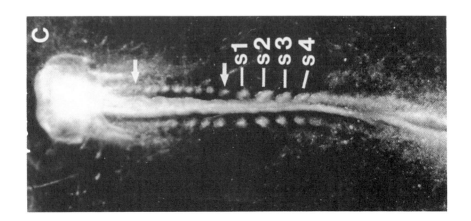

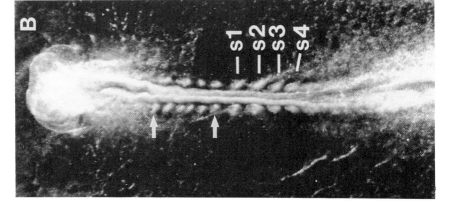

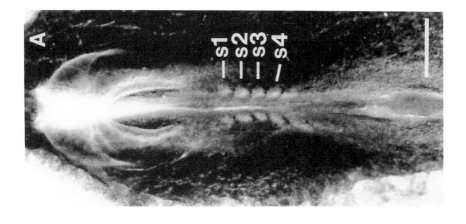

mesoderm. During normal development, a single, small somite will occasionally form just anterior to the first somite (82). This observation also suggests that even normally, the anterior mesoderm has a limited capacity to generate somites. In the case of the anterior somite-like structures of the chick, it remains to be determined whether their appearance in any way results from the effect of retinoic acid on the expression of homeobox genes. Homeotic-type transformations of vertebral identity can be caused by treatment of mouse embryos with retinoic acid (83) and it is possible that the anterior somitic structures shown in Figs. 6B and 6C are due to a related phenomenon.

RETINOID METABOLISM IN EARLY CHICK EMBRYOS

Induction of Ghox 2.9 by retinoic acid in ectopic position does not necessarily mean that retinoic acid is normally involved in the regulation of the normal patter of Ghox 2.9 expression. This would require a demonstration that primitive streak-stage embryos contain endogenous retinoic acid, a difficult analysis because of the small size of such early embryos. An alternative approach requiring less tissue is to demonstrate that primitive streak embryos can synthesize retinoic acid from its precursors retinol and retina (Fig. 1). Figure 7 shows the result of an experiment in which cytosol of Hamburger–Hamilton stage 4/5 embryos was prepared by differential centrifugation and incubated with retinal, the immediate precursor of retinoic acid (Fig. 1). In the chromatogram of Fig. 7 there is a peak at the elution position of authentic retinoic acid, strongly suggesting that embryos can convert retinol to retinoic acid. The reaction is NAD^+-dependent and synthesis occurs at a rate of approximately 112 pmol/mg protein $\times$ min, a range similar to that found in limb buds (R.J. and G.E., in preparation). The capability of synthesizing retinoic acid is about 10 times higher in embryonic tissue (*area pellucida*) than in the surrounding extraembryonic tissue (*area opaca*) of the same developmental stage.

Fig. 6. Induction of somite-like structures in anterior lateral plate mesoderm by treatment with retinoic acid. Stage 4 embryos were cultured with (or without) 6 μM all-*trans*-retinoic acid exactly as in Fig. 3, then grown in the absence of retinoic acid until the five-somite stage (Hamburger–Hamilton stage 8+). (A) Control embryo, cultured without retinoic acid. (B, C) Embryos cultured from stage 4 to 5 in the presence of retinoic acid. Both B and C have received the same treatment. Arrows indicate anterior and posterior extent of ectopic somite-like structures in the lateral plate mesoderm. s1–s5, normal somites. Bar: 400 μm.

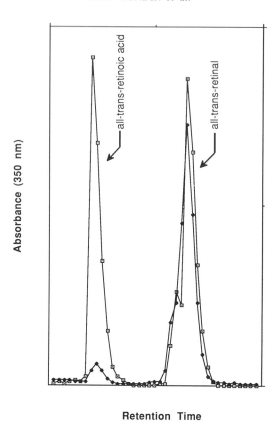

Retention Time

Fig. 7. Sample chromatogram showing the conversion of all-*trans*-retinal to all-*trans*-retinoic acid by the cytosolic protein from stage 4/5 chick embryos. In an *in vitro* assay, 15 μg cytosolic protein from either the *area pellucida* (embryo, open squares) or the *area opaca* (extraembryonic tissue, filled squares) were incubated with 1 μM all-*trans*-retinal in the presence of 2 mM NAD$^+$ at pH 8.5. The rate of retinoic acid synthesis in the *area pellucida* was at least 10 times higher than that in the surrounding *area opaca*.

In conclusion, there is evidence that retinoic acid and related molecules are actively synthesized by the embryo during the period when the main body axes are established. Mechanisms are also known whereby retinoids can regulate specific target genes by means of specific nuclear receptors. The expression of homeobox genes, key players in the process of morphogenesis, can be induced by exposing cultured cells or whole embryos to retinoic acid. Taken together, these studies indicate that both opportunity and mechanisms exist for endogenous retinoids to play an essential role during normal morphogenesis. It remains to be determined how the levels of endogenous retinoids are regulated within the embryo and, in mechanistic detail, how retinoids are used as signaling molecules during vertebrate development.

ACKNOWLEDGMENTS

This work was chiefly funded by grants from the American Cancer Society and the National Institutes of Health with additional support from the McKnight Endowment Fund for Neuroscience.

REFERENCES

1. Gurdon, J. B. (1992). *Cell* **68**, 185–199.
2. Jessell, T. M., and Melton, D. A. (1992). *Cell* **68**, 257–270.
3. McGinnis, W., and Krumlauf, R. (1992). *Cell* **68**, 283–302.
4. Krämer, H., Cagan, R. L., and Zipursky, S. L. (1991). *Nature* **352**, 207–212.
5. Crick, F. H. C. (1970). *Nature* **225**, 420–422.
6. Green, S., and Chambon, P. (1988). *Trends Genet.* **4**, 309–314.
7. Evans, R. M. (1988). *Science* **240**, 889–895.
8. Beato, M. (1989). *Cell* **56**, 335–344.
9. Laudet, V., Hänni, C., Coll, J., Catzeflis, F., and Stéhelin, D. (1992). *EMBO J.* **11**, 1003–1013.
10. Giguère, V., Ong, E. S., Segui, P., and Evans, R. M. (1987). *Nature* **330**, 624–629.
11. Petkovich, M., Brand, N. J., Krust, A., and Chambon, P. (1987). *Nature* **330**, 444–450.
12. Brand, N., Petkovich, M., Krust, A., Chambon, P., de The, H., Marchio, A., Tiollais, P., and Dejean, A. (1988). *Nature* **332**, 850–853.
13. Krust, A., Kastner, P., Petkovich, M., Zelent, A., and Chambon, P. (1989). *Proc. Natl. Acad. Sci. U.S.A.* **86**, 5310–5314.
14. Zelent, A., Krust, A., Petkovich, M., Kastner, P., and Chambon, P. (1989). *Nature* **339**, 714–717.
15. Benbrook, D., Lernhardt, E., and Pfahl, M. (1988). *Nature* **333**, 669–672.
16. Kastner, P., Krust, A., Mendelsohn, C., Garnier, J. M., Zelent, A., Leroy, P., Staub, A., and Chambon, P. (1990). *Proc. Natl. Acad. Sci. U.S.A.* **87**, 2700–2704.
17. Leroy, P., Krust, A., Zelent, A., Mendelsohn, C., Garnier, J. M., Kastner, P., Dierich, A., and Chambon, P. (1990). *EMBO J.* **10**, 59–69.
18. Giguère, V., Shago, M., Zirngibl, R., Tate, P., Rossant, J., and Varmuza, S. (1990). *Mol. Cell. Biol.* **10**, 2335–2340.
19. Hamada, L., Gleason, S. L., Levi, B. Z., Hirschfeld, S., Appella, E., and Ozato, K. (1989). *Proc. Natl. Acad. Sci. U.S.A.* **86**, 8289–8293.
20. Mangelsdorf, D. J., Ong, E. S, Dyck, J. A., and Evans, R. M. (1990). *Nature* **345**, 224–229.

21. Yu, V. C., Delsert, C., Andersen, B., Holloway, J. M., Devary, O. V., Näär, A. M., Kim, S. Y., Boutin, L.-M., Glass, C. K., and Rosenfeld, M. G. (1991). *Cell* **67**, 1251–1266.

22. Mangeldorf, D. J., Borgmeyer, U., Heyman, R. A., Zou, J. Y., Ong, E. S., Oro, A. E., Kakizuka, A., and Evans, R. M. (1992). *Genes Dev.* **6**, 329–344.

23. Leid, M., Kastner, P., Lyons, R., Nakshatri, H., Saunders, M., Zacharewski, T., Chen, J.-Y., Staub, A., Garnier, J.-M., Mader, S., and Chambon, P. (1992). *Cell* **68**, 377–395.

24. Heyman, R. A., Mangelsdorf, D. J., Dyck, J. A., Stein, R. B., Eichele, G., Evans, R. M., and Thaller, C. (1992). *Cell* **68**, 397–406.

25. Levin, A. A., Sturzenbecker, L. J., Kazmer, S., Bosakowski, T., Huselton, C., Allenby, G., Speck, J., Kratzeisen, C., Rosenberger, M., Lovey, A., and Grippo, J. F. (1992). *Nature* **355**, 359–361.

26. Zhang, X., Hoffmann, B., Tran, P. B.-V., Graupner, G., and Pfahl, M. (1992). *Nature* **355**, 441–446.

27. Kliewer, S. A., Umesono, K., Mangelsdorf, D. J., and Evans, R. M. (1992). *Nature* **355**, 446–449.

28. Dollé, P., Ruberte, E., Kastner, P., Petkovich, M., Stoner, C. M., Gudas, L. J., and Chambon, P. (1989). *Nature* **342**, 702–705.

29. Noji, S., Yamaai, T., Koyama, E., Nohno, T., and Taniguchi, S. (1989). *FEBS* **264**, 93–96.

30. Osumi-Yamashita, N., Noji, S., Nohno, T., Koyama, E., Doi, H., Eto, K., and Taniguchi, S. (1990). *FEBS* **264**, 71—74.

31. Ruberte, E., Dollé, P., Krust, A., Zelent, A., Morriss-Kay, G., and Chambon, P. (1990). *Development* **108**, 213–222.

32. Ruberte, E., Dollé, P., Krust, A., Chambon, P., and Morriss-Kay, G. (1991). *Development* **111**, 45–60.

33. Smith, S. M., and Eichele, G. (1991). *Development* **111**, 245–252.

34. Rowe, A., Richman, J. M., and Brickell, P. M. (1991). *Development* **111**, 1007–1016.

35. Noij, S., Nohno, T., Koyama, E., Muto, K., Ohyama, K., Aoki, Y., Tamura, K., Ohsugi, K., Ide, H., Taniguchi, S., and Saito, T. (1991). *Nature* **350**, 83–86.

36. Ellinger-Ziegelbauer, H., and Dreyer, C. (1991). *Genes Dev.* **5**, 94–104.

37. Mendelson, C., Ruberte, E., LeMeur, M., Morriss-Kay, G., and Chambon, P. (1991). *Development* **113**, 723–734.

38. Rowe, A., Eager, N. S. C., and Brickell, P. M. (1991). *Development* **111**, 771–778.

39. Osmond, M. K., Butler, A. J., Voon, F. C. T., and Bellairs, R. (1991). *Development* **113**, 1405–1417.

40. Thaller C., and Eichele G. (1987). *Nature* **327**, 625–628.

41. Durston, A. J., Timmermans, J. P. M., Hage, W. J., Hendriks, H. F. J., de Vries N. J., Heideveld, M., and Nieuwkoop, P. D. (1989). *Nature* **340**, 140–144.

42. Thaller, C., and Eichele, G. (1990). *Nature* **345**, 815–819.

43. Eichele, G., and Thaller, C. (1988). *Development* **103**, 473–483.

44. Duester, G. (1991). *In* "Drug and Alcohol Abuse Reviews" (R. R. Watson, ed.), Vol. 2, pp. 375–402. Humana Press, Clifton, NJ.

45. Roberts, A. B., Nichols, M. D., Newton, D. L., and Sporn, M. B. (1979). *J. Biol. Chem.* **254**, 6296–6302.

46. Wagner, M., Thaller, C., Jessell, T., and Eichele, G. (1990). *Nature* **345**, 819–822.

47. Jessell, T. M., Bovolenta, P., Placzek, M., Tessier-Lavigne, M., and Dodd, J. (1989). *Ciba Found. Symp.* **144**, 255–280.

48. Blomhoff, R., Green, M. H., Berg, T., and Norum, K. R. (1990). *Science* **250**, 399–404.
49. Perez-Castro, A. V., Toth-Rogler, L. E., Wei, L-N., and Nguyen-Hou, M. C. (1989). *Proc. Natl. Acad. Sci. U.S.A.* **86**, 8813–8817.
50. Vaessen, M-J., Meijers, J. H. C., Bootsma, D., and van Kessel, A. G. (1990). *Development* **110**, 371–378.
51. Dencker, L., Annerwallee, E., Busch, C., and Eriksson, U. (1990). *Development* **110**, 343–352.
52. Maden, M., Hunt, P., Eriksson, U., Kuroiwa, A., Krumlauf, R., and Summerbell, D. (1991). *Development* **111**, 35–44.
53. Lumsden, A., and Keynes, R. (1989). *Nature* **337**, 424–428.
54. Wilkinson, D. G., Bhatt, S., Cook, M., Boncinelli, E., and Krumlauf, R. (1989). *Nature* **341**, 405–409.
55. Strickland, S., and Mahdavi, V. (1978). *Cell* **15**, 393–403.
56. Boylan, J. F., and Gudas, L. J. (1991). *J. Cell Biol.* **112**, 965–979.
57. Ong, D. E., MacDonald, P. N., and Gubitosi, A. M. (1988). *J. Biol. Chem.* **263**, 5789–5796.
58. Chytil, F., and ul-Haq, R. (1990). *Eukaryotic Gene Expression* **1**, 61–73.
59. Näär, A. M., Boutin, J-M., Lipkin, S. M., Yu, V. C., Holloway, J. M., Glass, C. K., and Rosenfeld, M. G. (1991). *Cell* **65**, 1267–1279.
60. Umesono, K., Murakami, K. K., Thompson, C. C., and Evans, R. M. (1991). *Cell* **65**, 1255–1266.
61. Mangelsdorf, D. J., Umesono, K., Kliewer, S. A., Borgmeyer, U., Ong, E. S., and Evans, R. M. (1991). *Cell* **66**, 555–561.
62. Wei, L-N., Blaner, W. S., Goodman, D. S., and Nguyen-Huu, M. C. (1989). *Mol. Endocrinol.* **1**, 526–534.
63. Giguère, V., Lyn, S., Yip, P., Siu, C. H., and Amin, S. (1990). *Proc. Natl. Acad. Sci. U.S.A.* **87**, 6233–6237.
64. Duester, G., Shean, M. L., McBride, M. S., and Stewart, M. J. (1991). *Mol. Cell. Biol.* **105**, 1917–1923.
65. McGinnis, W., and Krumlauf, R. (1992). *Cell* **68**, 283–302.
66. Dony, C., and Gruss, P. (1988). *Differentiation* **37**, 115–122.
67. Baron, A., Featherstone, M. S., Hill, R. E., Hall, A., Galliot,, B., and Duboule, D. (1987). *EMBO J.* **6**, 2977–2986.
68. Mavilio, F., Simeone, A., Boncinelli, E., and Andrews, P. W. (1988). *Differentiation* **37**, 73–79.
69. Simeone, A., Acampora, D., Arcioni, L., Andrew, P. W., Boncinelli, E., and Mavilio, F. (1990). *Nature* **346**, 763–766.
70. La Rosa, G. J., and Gudas, L. J. (1988). *Mol. Cell. Biol.* **8**, 3906–3917.
71. Papalopulu, N., Lovell-Badge, R., and Krumlauf, R. (1991). *Nucleic Acid. Res.* **19**, 5497–5506.
72. Simeone, A., Acampora, D., Nigro, V., Faiella, A., D'Esposito, M., Stornaiuolo, A., Mavilio, F., and Boncinelli, E. (1991). *Mech. Dev.* **33**, 215–228.
73. Sive, H. L., Draper, B. W., Harland, R. L., and Weintraub, H. (1990). *Genes Dev.* **4**, 932–942.
74. Sundin, O. H., and Eichele, G. (1992). *Development* **114**, 841–852.
75. Ruiz i Altaba, A., and Jessell, T. M. (1991). *Genes Dev.* **5**, 175–187.
76. Ruiz i Altaba, A., and Jessell, T. M. (1991). *Development* **112**, 945–958.
77. Sharpe, C. R. (1991). *Neuron* **7**, 239–247.

78. Sive, H. L., and Cheng, P. F. (1991). *Genes Dev.* **5**, 1321–1332.
79. Morriss-Kay, G. M., Murphy, P., Hill, R. E., and Davidson, D. R. (1991). *EMBO J.* **10**, 2985–2995.
80. Sundin, O. H., and Eichele, G. (1990). *Genes Dev.* **4**, 1267–1276.
81. Meier, S. (1981). *Dev. Biol.* **83**, 49–61.
82. Hinsch, G. W., and Hamilton, H. L. (1956). *Anat. Rec.* **125**, 225–245.
83. Kessel, M., and Gruss, P. (1991). *Cell* **67**, 89–104.

PART V

PATTERN FORMATION

12

The Relationship between *Krox-20* Gene Expression and the Segmentation of the Vertebrate Hindbrain

M. ANGELA NIETO, LEILA C. BRADLEY,
AND DAVID G. WILKINSON
Laboratory of Eukaryotic Molecular Genetics
National Institute for Medical Research
The Ridgeway, Mill Hill, London NW7 1AA, United Kingdom

INTRODUCTION

Recent studies have shed light on cellular mechanisms that underlie pattern formation in a specific region of the developing central nervous system, the hindbrain. A series of bulges, the rhombomeres, are formed in the hindbrain neural epithelium in all vertebrate embryos, and studies of the chick have shown that they reflect a process of segmentation. Rhomobomeric sulci appear in a fixed order in the early neural epithelium (1), and upon forming, a restriction of cell movement across each morphological boundary occurs (2). The hindbrain thus becomes subdivided into a series of at least five compartments, corresponding to r2–6. This segmentation of the hindbrain neural epithelium correlates with, and presumably underlies, the subsequent development of neurons in rhombomere-restricted patterns; neurogenesis is first initiated in alternating rhombomeres, r2, r4, and r6; the Vth, VIIth, and IXth branchial motor nerves arise from adjacent pairs of rhombomeres, r2/r3, r4/r5, and r6/r7 (3).

Little is known of the genetic basis of the formation of rhombomeres and the specification of their phenotype. In *Drosophila* development certain of the key genes involved in segmentation and segment iden-

Cell–Cell Signaling in
Vertebrate Development

tity are expressed in segment-restricted patterns, and it is likely that a similar situation would occur in vertebrates. Indeed, several genes have been found to have rhombomere-restricted expression in the developing mouse embryo. The *Hox*-2.6, -2.7, and -2.8 homeobox genes have anterior limits of expression at the r6/7, r4/5, and r2/3 boundaries, respectively and thus pairs of rhombomeres express particular combinations of these genes (4). *Hox*-2.9 expression occurs in a different pattern from these, being restricted to a single rhombomere, r4, in the 9.5-day-old mouse embryo (4–6) and in the chick (7). The segmental expression of these genes suggests that they may have an analogous role to their *Drosophila* counterparts in the specification of segment identity.

We have fewer clues regarding the role of another putative regulatory gene with segmental expression in the hindbrain, the zinc-finger gene *Krox-20*. *Krox-20* was first identified as a gene whose transcription is rapidly up-regulated on treating quiescent fibroblasts with serum (8). This response occurs in the presence of cycloheximide and is thus primary and presumably involves the activation of the *Krox-20* gene through signal transduction pathways. *Krox-20* protein binds DNA in a sequence-specific manner and several lines of evidence suggest that it may act as a transcription factor (9,10). *In situ* hybridization studies showed that *Krox-20* expression occurs in two alternating rhombomeres, r3 and r5, in the 9.5-day-old mouse hindbrain (11), a pattern that correlates with several other features of hindbrain development that also exhibit a two-segment periodicity. Reticular and branchial motor nerves differentiate in r2, r4, and r6 prior to r3 and r5 (3) and r3 and r5 are unique in not forming migratory neural crest (12). In addition, grafting experiments suggest that an alternation in cellular properties underlies the formation of rhombomere boundaries; the juxtaposition of r3 and r5 does not lead to boundary formation, but a boundary is generated when either of these are grafted adjacent to any of the even-numbered rhombomeres (13). Finally, branchial motor nerve each arise from two adjacent rhombomeres (3) and pairs of rhombomeres express particular combinations of *Hox-2* genes (4).

CONSERVED EXPRESSION PATTERN OF *KROX-20*

It is likely that the presence of rhombomeres in all vertebrate species is matched by a conservation of the expression domains of genes with roles in the segmental development of the hindbrain. We have therefore cloned *Krox-20* sequences from chick and *Xenopus* and used

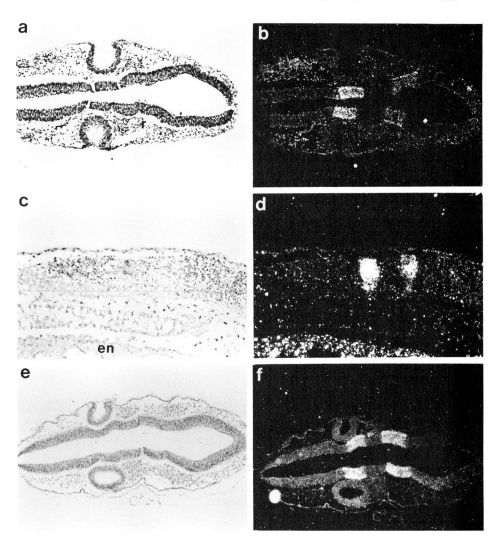

Fig. 1. Conserved patterns of *Krox-20* expression in the hindbrain of mouse, *Xeno-pus*, and chick embryos. *In situ* hybridization was carried out using appropriate homol-ogous *Krox-20* probes as described (16). (a,b) 9.5-day-old mouse embryo; (c,d) stage 28 *Xenopus* embryo; (e,f) stage 15 chick embryo. The apparent signal in the endoderm (en) of the *Xenopus* embryo is due to the refraction of light by yolky cells, not the hybridiza-tion of probe. Anterior is to the right in all photographs. Modified from Ref. (14).

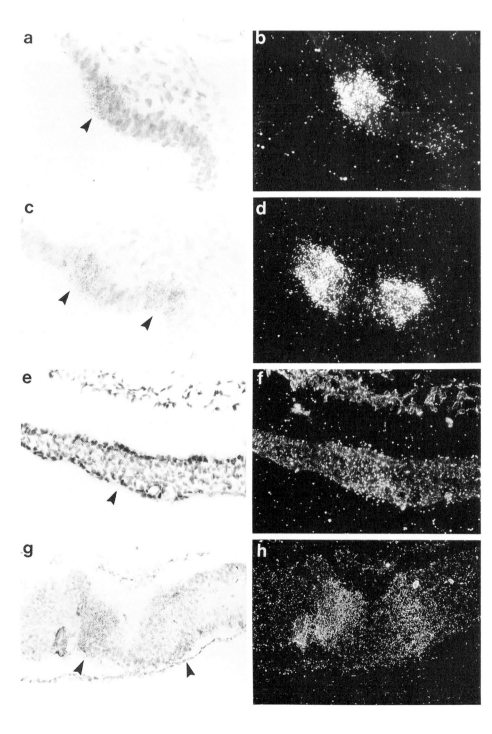

these to analyze the expression of *Krox-20* in these species (14, 15). *Krox-20* is expressed in two domains in the hindbrain in both *Xenopus* and the chick (Fig. 1). We cannot at present correlate the expression domains with rhombomeres in *Xenopus*, as the rhombomeric bulges form at later stages of development. However, in the chick, the domains of expression are seen to correspond to r3 and r5, the same pattern as in the mouse. These data indicate, at the very least, that the alternating expression of *Krox-20* in the early hindbrain is conserved between mammals, birds, and amphibians and support the idea that expression occurs in r3 and r5 throughout these vertebrate classes.

ESTABLISHMENT OF *KROX-20* EXPRESSION IN THE NEURAL EPITHELIUM

The two domains of *Krox-20* expression are established before rhombomeres appear at the morphological level in the mouse (4,11). At 8 days of development *Krox-20* is expressed in a single domain, and by 8.5 days a second, more posterior, domain has appeared (Fig. 2). These data are consistent with *Krox-20* being involved in early aspects of pattern formation, but since little is known of the cellular events of hindbrain segmentation in the mouse, these data do not indicate the timing of *Krox-20* expression relative to, for example, the establishment of lineage restriction. We have therefore analyzed *Krox-20* expression in the better characterized system, the chick embryo (14).

A single domain of *Krox-20* expression is detected in the chick hindbrain at the three-somite stage, and a second, more posterior, stripe of expression is detected in the seven-somite embryo (Fig. 2). The order of establishment of these domains of expression correlates with the formation of r3 prior to r5 (1), but occurs well before rhombomeres have formed. The anterior domain of *Krox-20* expression is detected >8 hr in advance of the formation of r3, and the posterior domain >3 hr prior to the formation of r5 (Fig. 3). In view of the spatial restriction of cell lineage occurring coincident with rhombomere boundary formation (2), these data show that *Krox-20* expression is initiated before compartments have formed.

Fig. 2.   Onset of *Krox-20* expression in the mouse and chick embryo. *In situ* hybridization analysis was carried out to examine the early stages of *Krox-20* expression. (a,b) 8-day-old mouse embryo; (c,d) 8.5-day-old mouse embryo; (e,f) three-somite chick embryo; (g,h) seven-somite chick embryo. The arrows indicate sites of *Krox-20* expression. (a–d) from Ref. (4) and (e–h) from Ref. (14).

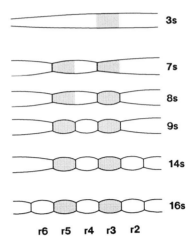

Fig. 3. The timing of *Krox-20* expression compared with rhombomere boundary formation. The diagram indicates the expression of *Krox-20* (shaded) in the developing chick hindbrain and the time course of rhombomere boundary formation (data from ref. 1). s, Somite stage; rf, rhombomere.

IMPLICATIONS FOR *KROX-20* FUNCTION

The conserved alternating expression of *Krox-20* in the mouse, chick, and *Xenopus* embryo suggests that this gene has a conserved role in hindbrain development. The timing of onset of expression indicates that hindbrain segmentation has been initiated by early neurula stages in all of these species. At present, it is only in the chick that we can compare the time of expression with the cellular events of hindbrain segmentation. Both the anterior and posterior stripe of *Krox-20* expression are established substantially in advance of the restriction of cell lineage at rhombomere boundaries. Moreover, the spatial pattern of expression correlates with the cellular properties shared by r3 and r5 that lead to boundary formation (13) and thus it is possible that *Krox-20* is involved in the formation of rhombomeres.

However, our data provide no direct evidence regarding gene function, and the early onset of *Krox-20* expression is also consistent with a role in other events of hindbrain segmentation. The expression in r3 and r5 correlates with the absence of neural crest migrating from these rhombomeres (12), the alternation in neuronal differentiation (3), and the expression of *Hox-2* genes with anterior limits at two segment intervals (4). By analogy with the situation in the early *Drosophila* embryo (reviewed in Ref. 17), the coupling of *Hox-2* gene expression to segment boundaries could occur through regulation by other transcription factors that are expressed in segment-restricted domains. Thus, *Krox-20* could, together with other genes, regulate the expression of *Hox-2.7* and *Hox-2.8* at high levels in r5 and r3–r5, respectively (4), and the restriction of *Hox-2.9* expression to r4 (4–7). The observation of *Krox-20* protein-binding sites in the *Hox-1.6* gene (9) provides provocative, though circumstantial evidence for a role of *Krox-20* in *Hox* gene regulation, and it will be very interesting to ascertain whether such sites exist in other *Hox* genes and to test their significance *in vivo*.

RELATIONSHIP BETWEEN *KROX-20* EXPRESSION AND LINEAGE RESTRICTION

Regardless of the function of *Krox-20*, once rhombomeres have formed expression is restricted to r3 and r5, and thus its early expression could indicate a commitment of cells to these rhombomeres. However, if such a cellular commitment is irrevocable, then it would be predicted that lineage restriction occurs coincident with, or prior to, the onset of expression, and not, as we have found, at later stages. There are two extreme models that explain these observations (Fig. 4). The initial *Krox-20* expression domains could be mosaic, consisting of a mixture of committed, expressing cells and uncommitted nonexpressing cells; clonal descendants of the former would be restricted to r3 and r5, whereas progeny of the latter cells could be subsequently recruited to either odd- or even-numbered rhombomeres. According to this model, the finding that some clones marked before rhombomere formation are restricted, whereas other clones are not (2), could be partly due to a mosaicism in cellular commitment and not only a consequence of whether clones have spread across prospective boundaries. A second possibility is that *Krox-20* expression is not a reliable marker of cell commitment and that it is down-regulated in cells that have migrated into prospective even-numbered rhombomeres. These models are not mutually exclusive, and it is possible that expression is both mosaic and not an indicator of a firm commitment to r3 and r5. As a first step toward addressing these possibilities, it will be important to analyze *Krox-20* expression at a single-cell resolution to ascertain whether expression might initially be mosaic. Such studies will provide a framework for future studies of the dynamics of the early events of hindbrain segmentation and a more direct analysis of the relationship between *Krox-20* expression and lineage restriction.

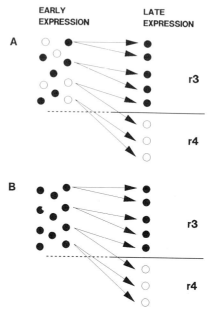

Fig. 4. Models of the relationship between *Krox-20* expression and lineage restriction. The diagrams depict two extreme models of how *Krox-20* expression might be related to lineage restriction. The expression pattern before lineage restriction in the region of prospective r3 is illustrated on the left, and the pattern after rhombomere formation on the right. *Krox-20* expressing cells are indicated by filled circles and nonexpressing cells by open circles. For each model the potential fate of four cells is indicated by arrows. According to (A) *Krox-20* expression is mosaic and expressing cells are committed to r3, whereas nonexpressing cells are uncommitted. According to (B) *Krox-20* expression is not mosaic, but expression does not correlate with a firm cellular commitment.

REFERENCES

1. Vaage, S. (1969). *Adv. Anat. Embryol. Cell Biol.* **41**, 1–88.
2. Fraser, S., Keynes, R., and Lumsden, A. (1990). *Nature* **344**, 431–435.
3. Lumsden, A., and Keynes, R. (1989). *Nature* **337**, 424–428.
4. Wilkinson, D. G., Bhatt, S., Cook, M., Boncinelli, E., and Krumlauf, R. (1989). *Nature* **341**, 405–409.
5. Murphy, P., Davidson, D. R., and Hill, R. E. (1989). *Nature* **341**, 156–159.
6. Murphy, P., and Hill, R. E. (1991). *Development* **111**, 61–74.
7. Sundin, O. H., and Eichele, G. (1990). *Genes Dev.* **4**, 1267–1276.
8. Chavrier, P., Zerial, M., Lemaire, P., Almendral, J., Bravo, R., and Charnay, P. (1988). *EMBO J.* **7**, 29–35.

9. Chavrier, P., Vesque, C., Galliot, B., Vigneron, M., Dolle, P., Duboule, D., and Charnay, P. (1990). *EMBO J.* **9,** 1209–1218.
10. Nardelli, J., Gibson, T. J., Vesque, C., and Charnay, P. (1991). *Nature* **349,** 175–178.
11. Wilkinson, D. G., Bhatt, S., Chavrier, P., Bravo, R., and Charnay, P. (1989). *Nature* **337,** 461–465.
12. Lumsden, A., Sprawson, N., and Graham, A. (1991). *Development* **113,** 1281–1291.
13. Guthrie, S., and Lumsden, A. (1991). *Development* **112,** 221–229.
14. Nieto, M. A., Bradley, L. C., and Wilkinson, D. G. (1991). *Development Suppl.* **2,** 59–62.
15. Bradley, L. C., Snape, A., Bhatt, S., and Wilkinson, D. G. (1993). *Mech. Dev.* **40,** 73–84.
16. Wilkinson, D. G., and Green, J. (1990). *In* "Post-implantation Mammalian Development" (A. Copp and D. Cockcroft, eds.), pp. 155–171. IRL Press, Oxford.
17. Ingham, P. W. (1988). *Nature* **335,** 25–34.

13

The Effect of Retinoids on Amphibian Limb Regeneration

JEREMY P. BROCKES
Ludwig Institute for Cancer Research
Middlesex Hospital/University College London Branch
London W1P 8BT, United Kingdom

INTRODUCTION

The aim of this chapter is to review the phenomena surrounding the effects of retinoic acid (RA) and other retinoids on amphibian limb regeneration. Regeneration of the urodele limb is one of the two key systems in which RA is able to respecify the positional value of cells participating in limb morphogenesis (1). Limb regeneration and development remain the most important cases for evaluating the morphogenetic effects of retinoids, in part because the resulting structures have a normal morphology, although they are misplaced on a particular axis. Although there is considerable interest in other retinoid effects on vertebrate development, for example, in the central nervous system (CNS) (2), it is much harder in these cases to distinguish developmental perturbations due to teratogenic effects from those which might involve morphogenetic respecification.

After describing the normal events of limb regeneration, I consider the position-dependent properties of the system before discussing their respecification by RA. A variety of experiments have clarified the nature of this effect at the cellular and tissue levels and these are reviewed in some detail, along with the evidence for contrasting effects of retinoids on the embryonic limb bud and the regeneration blastema. Although the effects on the proximodistal (P–D) axis are most marked in normal limb regeneration, it is possible to identify effects on the transverse axes in certain circumstances. It is important

Cell–Cell Signaling in
Vertebrate Development

to consider these experiments because of their implications for how retinoids act.

EVENTS OF LIMB REGENERATION

Amphibian limb regeneration is an example of epimorphic regeneration (3) that proceeds by formation of a growth zone or blastema at the plane of amputation. The blastemal cells are derived locally from the mesenchymal tissues of the limb stump, a process referred to as dedifferentiation (4). They proliferate in the environment at the end of the limb and undergo differentiation and morphogenesis to reconstruct the mesenchymal elements of the regenerate. The end of the limb is covered by the wound epidermis, a cell group that exerts an important influence on outgrowth of the underlying blastemal cells. The mechanisms involved in the formation of blastemal cells are not fully understood (5) and fall somewhat outside the scope of this account. It is important to recognize that the blastema, once generated, shows considerable morphogenetic autonomy and can be grafted to another location such as the eye socket or the fin where it gives rise to those structures characteristic of its origin (6). The use of monoclonal antibodies as cell markers has shown that blastemal cells are distinct from the cells of a normal limb and also distinct from the mesenchymal cells of the embryonic limb bud. Many of the distinctive aspects of regeneration reflect the properties of these remarkable cells, and one example is the phenomenon of distal transformation.

If the limb is amputated at any point along the P–D axis (shoulder to fingertip) the blastema exhibits a striking example of position-dependent identity in that it gives rise just to the missing structures. Blastemal cells inherit or otherwise derive from their location some property that leads to the generation of distal but never proximal regenerate tissues. Thus a wrist blastema regenerates a hand, while a shoulder blastema produces an arm. What happens therefore, when a wrist blastema is transplanted to a shoulder stump (Fig. 1, steps 6 and 7)? The result is a normal regenerate limb, but if the wrist blastema is marked, it is observed that the tissue up to the wrist is contributed by the shoulder stump, whereas the hand is derived from the wrist blastema (6). Thus the juxtaposition of proximal and distal tissue mobilizes the proximal partner to undergo regulated growth until the level of the distal one is reached. The phenomenon is an example of intercalary regeneration or intercalation, and P–D intercalation provides a second index of specification in the P–D axis. The end point is a normal limb

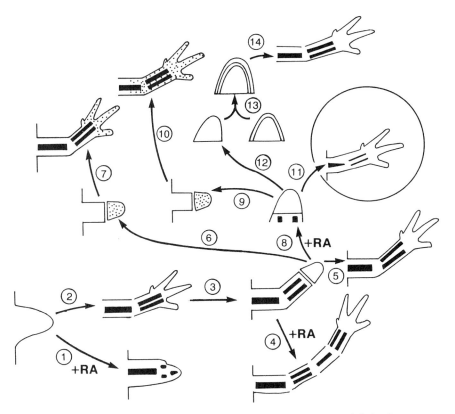

Fig. 1.   Summary of experiments analyzing the effects of RA on urodele limb regeneration. If a developing limb bud is exposed to RA (1), the predominant effect is to provoke skeletal deletions or truncations, giving hypomorphic limbs. After normal development (2), the limb can be amputated at the wrist (3) to yield a wrist blastema. If such a blastema is treated with an appropriate dose of RA, it gives rise (4) to a humerus, radius/ulna, and hand, whereas the normal result is a hand (5). If a wrist blastema is transplanted to a shoulder (6), a normal regenerate forms (7), but only the hand is contributed by the wrist blastema (speckled). If the wrist blastema is exposed to RA (8) and then transplanted to the shoulder (9), the blastema now contributes (10) more proximal structures to the regenerate. If the RA-treated blastema is transplanted to the orbit (11), it procedes to give a humerus, radius/ulna, and hand. After separating the blastema (12) into epithelial and mesenchyme fragments, the RA-treated mesenchyme can be recombined (13) with a normal mesenchyme and will give rise (14) to structures proximal to its point of origin.

and the extent of contribution of the two tissue sources is a measure of their relative disparity. The molecular basis of such specification is not understood but it is of great interest that both are reset by application of exogenous retinoids.

EFFECTS OF RETINOIDS ON LIMB REGENERATION

The effects of retinoids on limb regeneration are observed after bath application or intraperitoneal injection of either RA, a precursor such as retinol (vitamin A), or a synthetic nonmetabolized compound such as arotinoid (TTNPB). If an early regenerate at the dedifferentiation stage is exposed to retinoid in this way, the most striking early effect is an inhibition of blastemal proliferation so that regeneration is retarded relative to controls. After the level of retinoid declines following metabolism, the blastema grows out and gives rise to additional structures proximal to its plane of origin. Thus a wrist blastema may produce an entire arm if treated with an optimal dose of RA (Fig. 1, step 4). No other class of molecule is currently known to produce this effect of proximalizing the blastema. The resulting regenerates are referred to as serial duplications in the PD axis. The degree of serial duplication is a function of the concentration of retinoid and the time of exposure, thus it is graded and dose-dependent. At longer times of administration the retinoid may block regeneration permanently.

If a wrist blastema is exposed to RA and then transplanted to a shoulder stump, the resulting intercalary regenerate is derived solely from the wrist component (Fig. 1, step 10) (7). This is consistent with proximalization of the distal blastema such that when juxtaposed to the shoulder tissue (Fig. 1, step 9), there is no longer a positional disparity. Thus both indices of positional specification on the P–D axis are reset by RA in a proximal direction. These remarkable effects raise first the possibility that RA or a related molecule may be used as an endogenous signal to establish positional identity. Thus a shoulder blastema could encounter a higher concentration of RA than a wrist blastema, or possibly be more responsive to RA. At present there is no evidence to indicate that blastemal cells encounter a physiologically significant concentration of RA during normal regeneration. A second question concerns the mechanism of action of RA when it proximalizes the blastema, since this could be a key to understanding the molecular basis of axial specification. At this point I review the main features of action at the cell and tissue level as any attempts to identify molecular mechanisms must be grounded in such information.

The morphogenetic effects of retinoids are exerted ultimately on the mesenchymal compartment of the blastema (8). Thus a retinoid-treated mesenchyme gives a serial duplication after it is recombined with a normal wound epithelium (Fig. 1, steps 12–14). It is nonetheless possible that the respecification of the blastemal mesenchyme might proceed through interactions with the epithelium. The effect on the blastema is exhibited even after transplantation to a remote location such as the anterior chamber of the eye (Fig. 1, step 11) (9). This reflects the morphogenetic autonomy of the blastema that was described above. If retinoids are applied to the regenerating limb at different stages, their effect in respecifying a distal blastema is exerted only during the early period, referred to as dedifferentiation, when the blastemal cells arise from the stump mesenchyme (9, 10). The detailed significance of this finding for our understanding of the overall mechanism is not clear, but it is an important point that retinoids will only act when applied at this stage.

A most important aspect of the action is that it is exerted on regenerating and not on developing limbs. If vitamin A (retinol) is applied to axolotl larvae at a stage when the limb buds are developing it has a teratogenic effect, provoking formation of hypomorphic limbs with skeletal truncations and deletions (Fig. 1, step 1). In such larvae the forelimbs develop in advance of the hindlimbs and it is possible to have a hindlimb bud at a stage comparable to that of a forelimb blastema in the same animal. After exposure to an appropriate dose of vitamin A, the limb bud responds with deletions or truncations, while the blastema gives the characteristic serial duplications (11). This result graphically illustrates that retinoid responsiveness differs between development and regeneration. Any attempt to account for the molecular mechanism of respecification must ultimately explain this distinction.

Although this account has concentrated on effects on the P–D axis there are circumstances in which retinoids will effect the anteroposterior (A–P) axis. Although such effects are generally not seen in normal limb regeneration in urodeles, regeneration in anuran tadpole limbs treated with vitamin A produces regenerates that are serially duplicated in the P–D axis while exhibiting mirror-image duplications in the A–P axis (12, 13). It is possible to demonstrate effects on the A–P axis in urodeles by surgically constructing limbs derived solely from anterior tissue in the distal section (14). Such limbs will not regenerate successfully after amputation because of a requirement for a complete set of axial positional values. After treatment with RA, such a set is generated by virtue of the ability to posteriorize the blastema and thus produce regenerates that are complete in the A–P axis as well as showing the familiar serial duplications in the P–D axis. It should be noted that such effects are unidirectional in as much as limb sections derived solely from posterior tissue do not regenerate after retinoid treatment. More recently it has been possible to demonstrate that similar effects are observed in the dorsoventral (D–V) axis such that

RA ventralizes the blastema (15). Thus the emphasis on the P–D axis arises because in normal regeneration the limb is truncated in this axis, and the missing distal values are replaced by distal transformation. RA is apparently able to intervene in these events and proximalize the blastemal cells.

CONCLUSIONS

There are two major possibilities at the cellular level for how RA might respecify the limb blastema. It could act on blastemal cells to directly control their positional value on an axis. Such a mechanism might but need not necessarily reflect the operation of some endogenous signaling mechanism that controls cellular identity in a concentration-dependent fashion. This could broadly be referred to as a morphogen model. A second possibility would be that RA is able to set the boundary of an axis, but that the axial respecification depends on interactions within the limits of the morphogenetic field as established by its boundaries. A model of this latter kind has been proposed to account for the effects of RA on the chick limb bud (16). It is suggested that topical application of RA induces the adjacent tissue to become a polarizing region that establishes the posterior boundary of the A–P axis. The new polarizing region would then act in some unspecified way that does not involve RA to produce a morphogenetic duplication.

The most obvious possibility for such a boundary model in limb regeneration would be if RA acted by controlling the extent of blastema formation. If a distal blastema is treated with RA, it could increase the normal extent of blastema formation, thus extending the representation to more proximal values. Such a model has many problems and uncertainties, although the histological evidence currently available is consistent with an effect on the extent of dedifferentiation. The model does underline that we do not understand at which level RA may be acting or which primary processes it might control. It nonetheless remains a significant opportunity to tackle the problem of positional identity in vertebrate limb morphogenesis. New opportunities for understanding the molecular basis of such effects have arisen from the identification of receptors for RA in the limb and limb blastema (17), and the ability to manipulate their expression in cultured blastemal cells prior to introducing such cells back into the regenerate (18, 19).

REFERENCES

1. Brockes, J. P. (1989). *Neuron* **2**, 1285–1294.
2. Durston, A. J., Timmermans, J. P. M., Hage, W. J., *et al.* (1989). *Nature* **340**, 140–144.
3. Morgan, T. H. (1901). "Regeneration" MacMillan, London.
4. Wallace, H. (1981). "Vertebrate Limb Regeneration" Wiley, Chichester.
5. Ferretti, P., and Brockes, J. P. (1991). *Glia* **4**, 214–224.
6. Stocum, D. L. (1984). *Differentiation* **27**, 13–28.
7. Stocum, D. L., and Crawford, K. (1987). *Biochem. Cell Biol.* **65**, 750–761.
8. Maden, M. (1984). *J. Exp. Zool.* **230**, 387–392.
9. Thoms, S. D., and Stocum, D. L. (1984). *Dev. Biol.* **103**, 319–328.
10. Maden, M., Keeble, S., and Cox, R. A. (1985). *Roux Arch. Dev. Biol.* **194**, 228–235.
11. Scadding, S. R., and Maden, M. (1986). *J. Embryol. Exp. Morphol.* **91**, 19–34.
12. Niazi, I. A., and Saxena, S. (1978). *Folia Biol. (Krakow)* **26**, 3–11.
13. Scadding, S. R., and Maden, M. (1986). *J. Embryol. Exp. Morphol.* **91**, 35–53.
14. Kim, W.-S., and Stocum, D. L. (1986). *Dev. Biol.* **114**, 170–179.
15. Ludolph, D. C., Cameron, J. A., and Stocum, D. L. (1990). *Dev. Biol.* **140**, 41–52.
16. Wanek, N., Gardiner, D. M., Muneoka, K., and Bryant, S. V. (1991). *Nature* **350**, 81–83.
17. Ragsdale, C. W., Petkovich, M., Gates, P. B., Chambon, P., and Brockes, J. P. (1989). *Nature* **341**, 654–657.
18. Ragsdale, C. W., and Brockes, J. P. (1991). *In* "Structure and Function of Hormone Nuclear Receptors" (M. G. Parker, ed.). Academic Press, London.
19. Ferretti, P., and Brockes, J. P. (1988). *J. Exp. Zool.* **247**, 77–91.

PART VI

TRANSCRIPTION FACTORS

14

Role of Transcription Factor GATA-1 in the Differentiation of Hemopoietic Cells

LARYSA PEVNY,* M. CELESTE SIMON,†
VIVETTE D'AGATI,‡ STUART H. ORKIN,†
AND FRANK COSTANTINI*
*Departments of * Genetics and Development and ‡ Pathology*
College of Physicians and Surgeons
Columbia University
New York, New York 10032
and
† Division of Hematology-Oncology, Children's Hospital
Department of Pediatrics, Harvard Medical School
and the Howard Hughes Medical Institute
Boston, Massachusetts 02115

INTRODUCTION

GATA-1 is a transcription factor containing two highly conserved zinc-fingers, which binds to the general consensus motif A/TGATAA/G (Evans and Felsenfeld, 1989; Tsai *et al.*, 1989). This factor has been implicated in the regulation of many genes expressed in erythroid cells, including α- and β-globins (Beaupain *et al.*, 1990; Brady *et al.*, 1989; Cox *et al.*, 1991; Evans *et al.*, 1988; Evans and Felsenfeld, 1989, 1991; Gong *et al.*, 1991; Hannon *et al.*, 1991; Martin *et al.*, 1989; Tsai *et al.*, 1989, 1991; Wall *et al.*, 1988; Zon *et al.*, 1991c), as well as several genes expressed in mast cells and megakaryocytes (e.g., Mignotte *et al.*, 1989a,b; Ravid *et al.*, 1991; Youssoufian *et al.*, 1990; Zon *et al.*, 1991a,c). Expression of the GATA-1 factor in differentiated blood cell types appears to be limited to these three lineages (Tsai *et al.*, 1989;

Cell–Cell Signaling in
Vertebrate Development

Martin *et al.*, 1990; Romeo *et al.*, 1990). Although it has been proposed that GATA-1 may play an important role in the differentiation of the erythroid lineage, the evidence has so far been largely indirect (Evans *et al.*, 1988; Orkin, 1990; Reitman and Felsenfeld, 1988; Martin *et al.*, 1989; Martin and Orkin, 1990; Tsai *et al.*, 1991).

The task of defining the role of GATA-1 was complicated by the discovery that this factor is a member of a small family of related DNA-binding proteins, all of which share highly conserved finger domains and bind to the same consensus motif (Dorfman *et al.*, 1992; Ho *et al.*, 1991; Joulin *et al.*, 1991; Ko *et al.*, 1991; Marine and Winoto, 1991; Wilson *et al.*, 1990; Yamamoto *et al.*, 1990; Zon *et al.*, 1991b). GATA-2 is expressed in a fairly wide variety of cell types including erythroid cells, while GATA-3 is expressed primarily in T-cells and in the brain. Thus, the possibility exists that the various GATA factors may be functionally redundant for the activation of certain genes in cell types where they are coexpressed.

To test the importance of GATA-1 directly for the differentiation of various hematopoietic cell types, we derived a murine embryonic stem (ES) cell line in which the sole, X-linked copy of the GATA-1 gene was disrupted (Pevny *et al.*, 1991) by gene targeting (Capecchi, 1989). ES cells are pluripotent embryonic cells which, when introduced into normal mouse embryos, can colonize somatic tissues as well as the developing germ line (Evans and Kaufman, 1981; Martin, 1981; Bradley *et al.*, 1984). Thus, the developmental consequences of any genetic modifications to the ES cells can be tested by using the mutant cells to produce chimeric animals, thus introducing the mutation into the germ line (Robertson *et al.*, 1986). By disrupting the GATA-1 gene in an ES cell line, and producing chimeric mice using the mutant ES cells, we have established that this transcription factor is required for the terminal differentiation of erythroid cells (Pevny *et al.*, 1991; Simon *et al.*, 1992) and have also examined its importance for other hemopoietic lineages (Simon *et al.*, 1992, and unpublished data).

RESULTS

The GATA-1 locus was targeted in CCE line ES cells (Pevny *et al.*, 1991) using the positive/negative selection strategy of Mansour *et al.* (1988). Six ES cell clones, which had undergone the desired homologous recombination event at the GATA-1 locus, were identified, and two of these clones with normal XY karyotypes were used for analysis. Because the GATA-1 gene is X-linked (Zon *et al.*, 1990), and the CCE line was derived from a male embryo, the mutant ES cells lacked a wild-type allele of the GATA-1 gene. This permitted us to examine the phenotypic effects of the mutation by testing the ability of the mutant ES cells to contribute to various tissues and cell lineages in chimeric embryos or mice (Pevny *et al.*, 1991), or by allowing the ES cells to undergo hemopoietic differentiation *in vitro* (Simon *et al.*, 1992) (see Fig. 1).

A variety of tissues from embryonic, fetal, and neonatal chimeras were analyzed, using a glucose-phosphate isomerase (GPI-1) polymor-

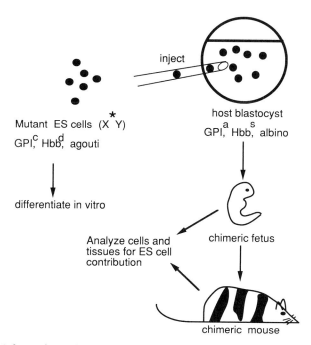

Fig. 1.   Scheme for analysis of the phenotypic effects of the GATA-1 mutation *in vivo* and *in vitro*. In one type of experiment, ES cells carrying the X-linked GATA-1 mutation, or wild-type ES cells, were microinjected into mouse blastocysts (Robertson, 1987), which were then implanted into pseudopregnant foster mothers and allowed to develop to fetal stages, or to term. Various tissues were dissected, and the percent ES cell contribution was determined based on the distribution of GPI-1 isozymes (Pevny *et al.*, 1991). Hemopoietic colonies were grown *in vitro* from chimeric Day 10 yolk sac, and the origin of different colony types was determined by GPI-1 analysis (unpublished data). In addition, the mutant ES cells, or wild-type ES cells, were allowed to form embryoid bodies *in vitro* (Deutschman *et al.*, 1985; Wiles and Keller, 1991), and differentiation of erythroid and other cell types was scored by cell morphology as well as by the presence of molecular markers, such as globin mRNAs (Simon *et al.*, 1992).

phism, as well as a hemoglobin polymorphism, to measure the fraction of each tissue derived from the ES cells. In fetal and neonatal mice, while all solid tissues examined showed extensive ES cell contributions, ranging from about 15% to over 50%, the whole blood contained very little (1 to 5%) or no detectable donor contribution, and the small ES cell-derived component was limited to the white cell fraction. Hemoglobin was exclusively of the host embryo type, with no ES cell-derived component. In contrast, control wild-type ES cells contributed extensively to the definitive red cell population. We concluded that the GATA-1 mutation impaired the ability of ES cell derivatives to form mature red blood cells of the definitive lineage, but permitted formation of at least a fraction of white blood cell types (Pevny *et al.*, 1991). Analysis of blood from Day 10 embryonic chimeras indicated that the formation of primitive, yolk sac-derived red blood cells was also blocked by the mutation (L.P. and F.C., unpublished data).

Among the chimeric fetuses produced with mutant ES cells, virtually all the highly chimeric ones were severely anemic or dead when examined at Day 14–16 of embryogenesis, while no anemia or death was observed among nonchimeric fetuses or chimeras made with wild-type ES cells. On histological examination, the chimeric fetal livers displayed a two- to threefold deficiency of maturing erythroid precursors, while no significant alteration in the number of myeloid cells was observed. Surprisingly, we observed a fivefold increase in the number of megakaryocytes in the chimeric fetal livers compared to those of control fetal livers. Consistent with the observation that extensive chimerism is usually lethal *in utero*, the chimeric mice that survived beyond birth showed much lower ES cell contributions than the fetal chimeras found *in utero*. While the cause of the anemia and death of chimeric fetuses has not been established, it appears that the host, wild-type hemopoietic cells are unable to effectively compensate for the erythopoietic defect of the mutant cells (Pevny *et al.*, 1991).

To examine in more detail the earliest stage at which erythroid differentiation is affected by the GATA-1 mutation, and to investigate the possible effects on other hemopoietic lineages, we have examined hemopoietic colonies grown *in vitro* from clonogenic precursors in the yolk sacs of Day 10 chimeric embryos. Individual colonies from methylcellulose cultures were picked, their origin (from ES cells vs. host embryo) determined by GPI-1 analysis of part of the colony, and the cell types present in each colony were determined by microscopic examination of Wright–Giemsa-stained preparations. While hemoglobinized erythroid colonies containing red cells of varying degrees of maturation arose from host-derived progenitors, no such colonies were found to derive from GATA-1 mutant progenitors. Instead, mutant erythroid and multipotential progenitors formed colonies in which erythroid differentiation appeared to be blocked at the proerythroblast stage, based on cellular morphology and staining. In contrast, the *in vitro* differentiation of macrophages, neutrophils, megakaryocytes, and mast cells appeared to be unaffected by the mutation (L.P. and F.C., unpublished data).

In a complementary approach, we have examined the ability of the mutant vs. wild-type ES cells to differentiate into hemopoietic lineages in a completely *in vitro* system. When cultured *in vitro* as embryoid bodies, under appropriate conditions ES cells can generate hemopoietic cells of erythroid and myeloid lineages (Doetschman *et al.*, 1985; Wiles and Keller, 1991). Differentiation of wild-type ES cells in this system resulted in the formation of clusters of hemoglobinized cells, including erythroid cells of varying degrees of maturation, and containing high levels of globin mRNAs. In contrast, the embryoid bodies formed by GATA-1 mutant ES cells failed to produce detectable globin mRNA and contained no erythroid cells more mature than the proerythroblast stage. Macrophages of similar morphology were produced by both the mutant and wild-type ES cell cultures. These results provided independent support for several of the conclusions drawn from the analysis of chimeric mice. In addition, the *in vitro* differentiation of ES cells represents a convenient system in which to assay the ability of wild-type or modified GATA-1 transgenes to complement the GATA-1 mutant phenotype (Simon *et al.*, 1992).

DISCUSSION AND CONCLUSIONS

We have disrupted the X-linked GATA-1 gene by targeted mutagenesis in embryonic stem cells and examined the phenotypic effects of the mutation by *in vivo* chimera analysis as well as by *in vitro* differentiation of the ES cells. These studies have established that the GATA-1 gene is required for the production of definitive red blood cells in fetal and adult chimeric mice (Pevny *et al.*, 1991) and for the production of primitive red blood cells in *in vitro* ES cell-derived embryoid bodies (Simon *et al.*, 1992) or in chimeric embryos (unpublished data). Several lines of evidence indicate that the absence of GATA-1 leads to a block in differentiation at the stage of the early erythroid precursor, or proerythroblast. First, more mature erythroblasts and normoblasts are underrepresented in chimeric fetal livers, suggesting that the mutant cells do not progress to these stages *in vivo* (Pevny *et al.*, 1991; unpublished data). Second, cells with the appearance of proerythroblasts accumulate in GATA-1 mutant colonies grown *in vitro* from hemopoietic tissues of chimeric fetuses (unpublished data), as well as in embryoid bodies derived from the mutant ES cells *in vitro* (Simon *et al.*, 1992). Third, the number of committed erythroid and mixed erythroid —myeloid progenitors in the yolk sac appeared not to be significantly affected by the mutation (unpublished data), suggesting that the mutation does not affect cells until a later point in the erythroid pathway. Finally, RNA analysis of the *in vitro*-derived embryoid bodies indicated that the production of detectable globin mRNA is prevented by the mutation, consistent with a block to differentiation at an early stage or erythroid cell maturation (Simon *et al.*, 1992).

These results are consistent with the evidence that GATA-1 is required for the activation of many genes whose products accumulate during the terminal stages of erythroid cell differentiation and suggest that the factor is not required for earlier events in the formation of to the erythroid lineage. Although GATA-1 has been implicated in the regulation of the erythropoietin receptor gene (Chiba *et al.*, 1991; Yousouffian *et al.*, 1990; Zon *et al.*, 1991c), it is not yet known whether expression of the Epo receptor is affected by the mutation.

The GATA-1 mutation did not interfere with the *in vitro* differentiation of macrophages or neutrophils, either in colonies derived from chimeric embryo yolk sac progenitors or in embryoid bodies derived directly from cultured ES cells. This is consistent with the absence of detectable GATA-1 expression in normal, mature cells of these lineages. Although GATA-1 expression has been detected in cell lines thought to represent multipotential erythroid–myelodi progenitors (Crotta *et al.*, 1990; Whitelaw *et al.*, 1990), our results indicate that the absence of GATA-1 does not block the formation of multipotential or committed progenitors, or their progression into the myeloid lineages.

The fact that GATA-1 is expressed in megakaryocytes and mast cells has suggested that this transcription factor may also be important for differentiation in these lineages, or perhaps for the development of a progenitor committed to the erythroid, megakaryocyte, and mast cell lineages (Martin *et al.*, 1990; Romeo *et al.*, 1990). However, mast cells and megakaryocytes have been observed in GATA-1 mutant yolk sac-derived *in vitro* colonies, and their numbers and cellular morphologies are similar to those seen in wild-type colonies (unpublished data). The possibility remains that GATA-1 may be required for subsequent differentiation events not easily observed *in vitro* (e.g., platelet formation or tissue mast cell formation); alternatively, other members of the

GATA family of transcription factors may be able to substitute for GATA-1 in these lineages.

ACKNOWLEDGMENTS

This work was supported by grants to F.C. and S.H.O. from the NIH and the ACS. L.P. is a march of Dimes Predoctoral Fellow; S.H.O. is an Investigator and M.C.S. is an Associate of the Howard Hughes Medical Institute.

REFERENCES

Beaupain, D., Eleouet, J. F., and Romeo, P. H. (1990). Initiation of transcription of the erythroid promoter of the porphobilinogen deaminase gene is regulated by a cis-acting sequence around the cap site. *Nucleic Acids Res.* **18**, 6509–6515.

Bradley, A., Evans, M., Kaufman, M. H., and Robertson, E. (1984). Formation of germ-line chimaeras from embryo-derived teratocarcinoma cell lines. *Nature* **309**, 255–256.

Brady, H. J., Sowden, J. C., Edwards, M., Lowe, N., and Butterworth, P. H. (1989). Multiple GF-1 binding sites flank the erythroid specific transcription unit of the human carbonic anhydrase I gene. *FEBS Lett.* **257**, 451–456.

Capecchi, M. R. (1989). The new mouse genetics: Altering the genome byt gene targeting. *Trends Genet.* **5**, 70–76.

Chiba, T., Ikawa, Y., and Todokoro, K. (1991). GATA-1 transactivates erythropoietin receptor gene, and erythropoietin receptor-mediated signals enhance GATA-1 gene expression. *Nucleic Acids Res.* **19**, 3843–3848.

Cox, T. C., Bawden, M. J., Martin, A., and May, B. K. (1991). Human erythroid 5-aminolevulinate synthase: Promoter analysis and identification of an iron-responsive element in the mRNA. *EMBO J.* **10**, 1891–1902.

Crotta, S., Nicolis, S., Ronchi, A., Ottolenghi, S., Ruzzi, L., Shimada, Y., Migliaccio, A. R., and Migliaccio, G. (1990). Progressive inactivation of the expression of an erythroid transcriptional factor in GM- and G-CSF-dependent myeloid cell lines. *Nucleic Acids Res.* **18**, 6863–6869.

Doetschman, T., Eisetter, H., Katz, M., Schmidt, W., and Kemler, R. (1985). The in vitro development of blastocyst-derived embryonic stem cell lines: Formation of vesceral yolk sac, blood islands, and myocardium. *J. Embryol. Exp. Morphol.* **87**, 27–45.

Dorfman, D. M., Wilson, D. B., Bruns, G. A. P., and Orkin, S. H. (1992). Human transcription factor GATA-2. *J. Biol. Chem.* **267**, 1279–1285.

Evans, M. J., and Kaufman, M. H. (1981). Establishment in culture of pluripotential cells from mouse embryos. *Nature* **292**, 154–156.

Evans, T., Reitman, M., and Felsenfeld, G. (1988). An erythrocyte-specific DNA-binding factor recognizes a regulatory sequence common to all chicken globin genes. *Proc. Natl. Acad. Sci. U.S.A.* **85**, 5976–5980.

Evans, T., and Felsenfeld, G. (1989). The erythroid-specific transcription factor Eryf1: A new finger protein. *Cell* **58**, 877–885.

Evans, T., and Felsenfeld, G. (1991). trans-Activation of a globin promoter in noneryth-roid cells. *Mol. Cell. Biol.* **11**, 843–853.

Gong, Q. H., Stern, J., and Dean, A. (1991). Transcriptional role of a conserved GATA-1 site in the human epsilon-globin gene promoter. *Mol. Cell. Biol.* **11**, 2558–2566.

Hannon, R., Evans, T., Felsenfeld, G., and Gould, H. (1991). Structure and promoter activity of the gene for erythroid transcription factor GATA-1. *Proc. Natl. Acad. Sci. U.S.A.* **88**, 3004–3008.

Ho, I. C., Vorhees, P., Marin, N., Oakley, B. K., Tsai, S. F., Orkin, S. H., and Leiden, J. M. (1991). Human GATA-3: A lineage-restricted transcription factor that regulates the expression of the T cell receptor alpha gene. *EMBO J.* **10**, 1187–1192.

Joulin, V., Bories, D., Eleouet, J. F., Labastie, M. C., Chretien, S., Mattei, M. G., and Romeo, P. H. (1991). A T-cell specific TCR delta DNA binding protein is a member of the human GATA family. *EMBO J.* **10**, 1809–1816.

Ko, L. J., Yamamoto, M., Leonard, M. W., George, K. M., Ting, P., and Engel, J. D. (1991). Murine and human T-lymphocyte GATA-3 factors mediate transcription through a cis-regulatory element within the human T-cell receptor delta gene enhancer. *Mol. Cell. Biol.* **11**, 2778–2784.

Mansour, S. L., Thomas, K. R., and Capecchi, M. R. (1988). Disruption of the proto-oncogene *int-2* in mouse embryo-derived stem cells: A general strategy for targeting mutations to non-selectable genes. *Nature* **236**, 438–452.

Marine, J., and Winoto, A. (1991). The human enhancer-binding protein Gata3 binds to several T-cell receptor regulatory elements. *Proc. Natl. Acad. Sci. U.S.A.* **88**, 7284–7288.

Martin, D. I. K., Tsai, S.-F., and Orkin, S. H. (1989). Increased γ-globin expression in a nondeletion HPFH mediated by an erythroid-specific DNA-binding factor. *Nature* **338**, 435–438.

Martin, D. I., and Orkin, S. H. (1990). Transcriptional activation and DNA binding by the erythroid factor GF-1/NF-E1/Eryf 1. *Genes Dev.* **4**, 1886–1898.

Martin, D. I. K., Zon, L. I., Mutter, G., and Orkin, S. H. (1990). Expression of an erythroid transcription factor in megakaryocytic and mast cell lineages. *Nature* **344**, 444–447.

Martin, G. R. (1981). Isolation of a pluripotential cell line from early mouse embryos cultured in medium conditioned by teratocarcinoma stem cells. *Proc. Natl. Acad. Sci. U.S.A.* **78**, 7634–7638.

Mignotte, V., Eleouet, J. F., Raich, N., and Romeo, P. H. (1989a). Cis- and trans-acting elements involved in the regulation of the erythroid promote of the human porphobilinogen deaminase gene. *Proc. Natl. Acad. Sci. U.S.A.* **86**, 6548–6552.

Mignotte, V., Wall, L., deBoer, E., Grosveld, F., and Romeo, P.-H. (1989b). Two tissue-specific factors bind the erythroid promoter of the human porphobilinogen deaminase gene. *Nucleic Acids Res.* **17**, 37–54.

Orkin, S. H. (1990). Cell-specific transcription and cell differentiation in the erythroid lineage. *Curr. Opin. Cell Biol.* **2**, 1003–1012.

Pevny, L., Simon, M. C., Robertson, E., Klein, W., Tsai, S.-F., D'Agati, V., Orkin, S. H., and Costantini, F. (1991). Erythroid differentiation in chimeric mice blocked by a targeted mutation in the gene for transcription factor GATA-1. *Nature* **349**, 257–260.

Ravid, K., Doi, T., Beeler, D. L., Kuter, D. J., and Rosenberg, R. D. (1991). Transcriptional regulation of the rat platelet factor 4 gene: Interaction between an enhancer/silencer domain and the GATA site. *Mol. Cell. Biol.* **11**, 6116–6127.

Reitman, M., and Felsenfeld, G. (1988). Mutational analysis of the chicken β-globin enhancer reveals two positive acting domains. *Proc. Natl. Acad. Sci. U.S.A.* **85**, 6267–6271.

Robertson, E., Bradley, A., Kuehn, M., and Evans, M. (1986). Germ-line transmission of genes introduced into cultured pluripotential cells by retroviral vector. *Nature* **323**, 445–448.

Robertson, E. J. (1987). *In* "Teratocarcinomas and Embryonic Stem Cells: A Practical Approach" (E. J. Robertson, ed.), pp. 71–112. IRL, Oxford.

Romeo, P.-H., Prandini, M.-H., Joulin, V., Mignotte, V., Prenant, M., Vainchenker, W., Marguerie, G., and Uzan, G. (1990). Megakaryocytic and erythrocytic lineages share specific transcription factors. *Nature* **344**, 447–449.

Simon, M. C., Pevny, L., Wiles, M. V., Keller, G., Costantini, F., and Orkin, S. H. (1992). Rescue of erythroid development in gene targeted GATA-1 mouse embryonic stem cells. *Nature Genet.* Vol 1, pp. 92–98.

Tsai, S.-F., Martin, D. I. K., Zon, A. D., D'Andrea, A. D., Wong, G. G., and Orkin, S. H. (1989). Cloning of cDNA for the major DNA binding protein of the erythroid lineage through expression in mammalian cells. *Nature* **339**, 446–451.

Tsai, S. F., Strauss, E., and Orkin, S. H. (1991). Functional analysis and in vivo footprinting implicate the erythroid transcription factor GATA-1 as a positive regulator of its own promoter. *Genes Dev.* **5**, 919–931.

Wall, L., deBoer, E., and Grosveld, E. (1988). The human β-globin gene 3' enhancer contains multiple binding sites for an erythroid-specific protein. *Genes. Dev.* **2**, 1089–1100.

Whitelaw, E., Tsai, S. F., Hogben, P., and Orkin, S. H. (1990). Regulated expression of globin chains and the erythroid transcription factor GATA-1 during erythropoiesis in the developing mouse. *Mol. Cell. Biol.* **10**, 6596–6606.

Wiles, M. V., and Keller, G. (1991). Multiple hematopoietic lineages develop from embryonic stem (ES) cells in culture. *Development* **111**, 259–267.

Wilson, D. B., Dorfman, D. M., and Orkin, S. H. (1990). A nonerythroid GATA-binding protein is required for function of the human preproendothelin-1 promoter in endothelial cells. *Mol. Cell. Biol.* **10**, 4854–4862.

Yamamoto, M., Ko, L. J., Leonard, M. W., Beug, H., Orkin, S. H., and Engel, J. D. (1990). Activity and tissue-specific expression of the transcription factor NF-E1 multigene family. *Genes Dev.* **4**, 1650–1662.

Youssoufian, H., Zon, L., Orkin, S. H., d'Andrea, A. D., and Lodish, H. F. (1990). Structure and transcription of the mouse erythropoietin receptor gene. *Mol. Cell. Biol.* **10**, 3675–3682.

Zon, L. I., Tsai, S.-F., Burgess, S., Matsudaira, P., Bruns, G. A. P., and Orkin, S. H. (1990). The major human erythroid DNA-binding protein (GF-1): Primary sequence and localization of the gene to the X chromosome. *Proc. Natl. Acad. Sci. U.S.A.* **87**, 668–672.

Zon, L. I., Gurish, M. F., Stevens, R. L., Mather, C., Reynolds, D. S., Austen, K. F., and Orkin, S. H. (1991a). GATA-binding transcription factors in mast cells regulate the promoter of the mast cell carboxypeptidase A gene. *J. Biol. Chem.* **266**, 22948–22953.

Zon, L. I., Mather, C., Burgess, S., Bolce, M. E., Harland, R. M., and Orkin, S. H. (1991b). Expression of GATA-binding proteins during embryonic development in Xenopus laevis. *Proc. Natl. Acad. Sci. U.S.A.* **88**, 10642–10646.

Zon, L. I., Youssoufian, H., Mather, C., Lodish, H. F., and Orkin, S. H. (1991c). Activation of the erythropoietin receptor promoter by transcription factor GATA-1. *Proc. Natl. Acad. Sci. U.S.A.* **88**, 10638–10641.

15

Expression of *Hox-2* Genes and Their Relationship to Regional Diversity in the Vertebrate Head

PAUL HUNT AND ROBB KRUMLAUF
Laboratory of Developmental Neurobiology
MRC National Institute for Medical Research
The Ridgeway, Mill Hill
London, England NW7 1AA

INTRODUCTION

A great deal of excitement in vertebrate embryology has come with the identification of growth factors, receptors, and transcription factors that are highly conserved between species. It has been a basic hope that this conservation would reflect common regulatory functions and provide a means of examining the underlying molecular mechanisms in developmental hierarchies for pattern formation. One family of nuclear transcription factors, the *Hox* homeobox-containing genes, has received particular attention because increasing evidence suggests that they play a conserved role in vertebrate embryogenesis in the mechanisms specifying regional diversity. The purpose of this paper is to review some of the basic properties of the *Hox* clusters, explore their patterns of expression and potential roles in patterning the branchial region of the head, and discuss these results in terms of interactions required for head development.

Cell–Cell Signaling in
Vertebrate Development

CONSERVATION AND PROPERTIES
OF THE *Hox* NETWORK

The vertebrate *Hox* family is a group of 35–40 genes, organized in four chromosomal clusters, which contain a homeobox motif related to those in the *Drosophila* ANT-C/BX-C (HOM-C) homeotic complexes (1–3). Sequence analysis of genes from a single cluster reveal that the adjacent genes were originally derived by tandem duplication and divergence from a progenitor homeobox gene in an unidentified primordial invertebrate (4). During vertebrate evolution the *Hox* clusters themselves were involved in widespread chromosomal duplication events which gave rise to the four *Hox* clusters seen in most vertebrate species. A consequence of the cluster duplications is that individual members of any cluster are highly related to specific members in another cluster, forming subfamilies or paralogous groups (1–3, 5–7). Based on the extensive homology between subfamily members in multiple regions of the proteins there is a distinct possibility for function overlap or redundancy between members of the *Hox* network.

Detailed structural comparisons reveal that the mouse (*Hox*) and *Drosophila* (*HOM-C*) homeotic complexes share many conserved features, suggesting that they are truly homologous clusters derived from a common ancestor and that they may have some conserved functional roles in development (1–3, 8). Figure 1 summarizes the organizational relationship between the mouse and *Drosophila* complexes and some aspects of their general expression patterns. One of the most striking features of both the *Hox* and *HOM-C* gene clusters is that there is a direct correlation between the position of a gene in the complex and its relative domain of expression along the embryonic axis, which is termed spatial collinearity (1, 2, 9, 10). Genes in the *Hox* clusters all have the same 5′–3′ orientation with respect to transcription. Members at the 5′ end are expressed in the most posterior domains and each successively more 3′ gene along a cluster has a more anterior boundary of expression. This type of spatial collinearity has been observed in the nervous system (1, 2, 8, 11–14), the limbs (15–17), and the mesoderm (14, 18, 19). In addition to the spatial correlations with gene organization, collinearity has also been observed with respect to the temporal order of their activation during development and to the response of *Hox* genes to retinoic acid (7, 20–23). These collinear relationships are highly conserved in all vertebrates suggesting that they are a common property of the *Hox* clusters and lead to the suggestion that the *Hox* complexes could act as a

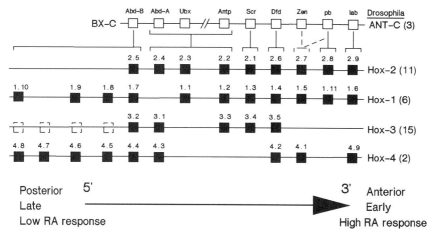

Fig. 1. Homologies in *HOM-C* and *Hox* homeobox complexes; alignment of the mouse (*Hox*) and *Drosophila* (*HOM-C*) homeobox gene clusters. The vertical rows of boxes indicates genes which are highly related to each other forming subfamilies. The brackets above the mouse genes identify the *Drosophila* homologue. Note that not all mouse *Hox* clusters have members in each subfamily. Solid boxes are sequenced genes and the dashed boxes genes present in the human clusters but not as yet identified in mouse. The large arrows beneath the clusters refer to the trends in collinear expression where successive genes from left to right have increasing anterior boundaries, earlier stages of expression, and sensitivity to retinoic acid.

molecular representation of different axial coordinates in the embryo (1, 2, 24).

The finely regulated patterns of spatial domains imply that the relative boundaries of *Hox* expression are important with respect to their functional roles. The vertebrate *Hox* proteins are believed to function as transcription factors involved in the specification of positional information along the anteroposterior axis, by analogy to their *Drosophila* counterparts. These restricted patterns of expression are thought to provide part of a basic molecular code, such that a particular combination of *Hox* genes together gives a positional address that can be used to generate differential structures. Experimental support for this hypothesis is derived from the ectopic expression of *Hox* genes in transgenic mice (19, 25), which result in new combinations of *Hox* gene products and the transformation or alteration of axial structures. Therefore, understanding the molecular basis of the transcriptional control of the *Hox* genes is important for examining their function and for identifying signals in the regulatory hierarchy.

Hox EXPRESSION IN THE BRANCHIAL ARCHES: A *Hox* CODE FOR THE HEAD

Expression before and during Crest Migration

In initial studies we found that *Hox-2* genes were segmentally expressed in the hindbrain, and we also wanted to determine if these patterns of expression extended to other regions of the head to explore the possibility that rhombomeric segmentation might play a general role in organization of the head. *Hox* genes are expressed in the cranial ganglia and the regions of the gut where the visceral nerve plexus is located, both of which are derived from cephalic neural crest (5, 8, 26). However, these early studies were carried out at stages of development when neural crest migration is complete and cell differentiation is well under way. If *Hox* genes are involved in the initial spatial patterning of neural crest structures as well as their differentiation, expression would be expected to occur in the crest as it emerges from the neural plate, as there is evidence that particular regions of neural plate have the potential to produce spatially specified crest before it emerges (27).

In mouse the neural crest begins to leave the margins of the neural plate at 8 days of development (4 somites) in the cranial region, and by 8.5 days (11 somites) migration is well under way (28, 29). *Hox-2.8* expression shows spatially restricted expression at the 0-somite stage within the hindbrain, as shown in Fig. 2A, and by 1 somite the anterior limit of its expression domain coincides with a stripe of *Krox-20* expression (Figs. 2B and 2C). These expression limits correspond to the boundary of what will be the future rhombomeres r2 and r3, although this stage is before the appearance of morphological features which allow precise positioning of expression domains. *Hox-2.9* and *Hox-1.6* have also reached their anterior limits in the neural tube at a similar stage (30). Thus *Hox-2.8* and *Hox-2.9* show spatially restricted expression within the hindbrain at a time before the first emergence of cranial crest.

To determine the timing of the onset of *Hox-2.8* expression in the crest-derived mesenchyme, we examined the expression in neural groove stage (8.5 day) mouse embryos in transverse section. A series of sections are shown in Fig. 3, where the plane is not quite perpendicular to the long axis of the embryo; thus, one side of the neural plate is more anterior than the other. Sections in the series more anterior than those shown here showed no labeling above background. In the most anterior section (Fig. 3A), only the neural plate is labeled and only on

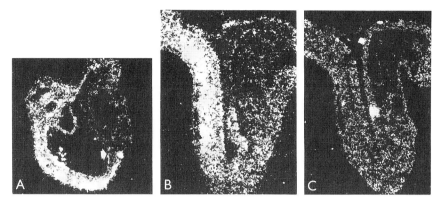

Fig. 2. Expression of *Hox-2.8* in 8-day-old mouse embryos. (A) Sagittal section of a presomitic mouse embryo, hybridized with *Hox-2.8*. The anterior boundary of expression is indicated by the white arrow. (B,C) Near adjacent sections of a 1-somite embryo (B) hybridized with *Hox 2.8* or (C) hybridized with *Krox-20*. The anterior limits of the *Hox-2.8* and *Krox-20* domains coincide.

one side. This section is at the level of the anterior limit of expression, between r2 (right-hand side, not expressing) and r3 (left-hand side, expressing). A near adjacent, more posterior section (Fig. 3B) shows labeling of both sides of the neural plate. Figure 3C shows an additional domain of expression lateral to the dorsal edge of the neural plate. This domain corresponds to migrating neural crest (nc) (29, 31–33) and is continuous with the neural plate. An interpretation of the rhombomere axial level of each section is indicated by the diagram on the left-hand side of the figure. It suggests that only two rhombomeres produce neural crest or correspond to areas of reduced crest emigration, in agreement with previous studies in chicken and rat embryos (34–36). Expression is not seen in noncrest mesenchyme underlying the neural plate. Thus it seems that where neural crest does arise it expresses *Hox-2* genes from time of emergence and that the crest migrating into the arches has a *Hox-2* label or code.

Expression after Cranial Crest Migration Is Complete

Figure 4 shows a series of coronal sections of a 9-day-old embryo hybridized with *Hox-2.8*. The position of the otocyst adjacent to rhombomeres 5 and 6 (O, Fig. 4A) allows orientation within the hindbrain. Expression is detected in the neural tube with a boundary at r2/3, and in the VII/VIII cranial ganglion complex (g), as previously

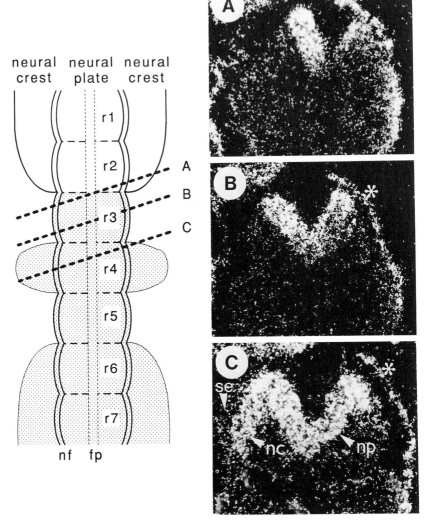

Fig. 3. *Hox-2* expression in migrating and premigratory neural crest; expression of *Hox-2.8* in serial, transverse sections of an 8.5-day-old mouse embryo hindbrain. The relative position of sections is shown in the diagram on the left-hand side, with A the most anterior section. The sections are slightly oblique, such that the right-hand side of each section is more anterior than the left-hand side. The asterisks indicate extraembryonic membranes that are expressing *Hox-2.8* and serve as a positive control for hybridization. The diagram on the left-hand side indicates which rhombomeres the sections are passing through and the plane of the section. Tissues known to be expressing *Hox-2.8* are shaded in gray stipple. The lobes lateral to the neural plate indicate areas of neural crest that are known to be produced by particular lengths of neural plate. nf, Neural fold; fp, floorplate; se, surface ectoderm; nc, migrating neural crest; np, neural plate.

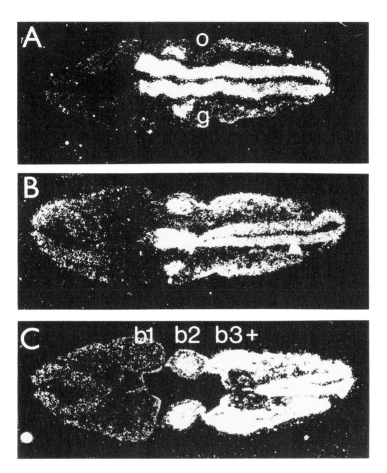

Fig. 4. *Hox-2.8* expression in the hindbrain and branchial arches after neural crest migration is complete. A set of serial sections coronal to the hindbrain is shown of a 9-day-old mouse embryo. (A) Most dorsal, showing expression in the hindbrain and the VII/VIII ganglion. (B) Ventral to (A), showing expression in the hindbrain and in addition two isolated patches of expression lateral of the neural tube. (C) More ventral than (B), passing through the branchial arches. The first arch does not express, while the second and more posterior arches express. Expression is seen in those parts of the head known to be colonized by neural crest. o, Otic placode; g, VII/VIII ganglion; b1–b3, first/third branchial arches.

described (8). In more ventral sections (Fig. 4B) it is possible to see the structures of the first and second branchial arches, and the boundary of *Hox-2.8* maps to the junction between the first and second arch. Hybridization is seen in the center of arch 2 and in a more posterior domain, as well as within the neural tube itself. In Fig. 4C, arch 2 is visible as a discrete structure, and it is clear that there is no expression in arch 1, while arch 2 and more posterior regions express *Hox-2.8* at high levels. *Hox-2.8* expression is confined to arch 2 and more posterior areas colonized by neural crest. This expression is in a region that also contains paraxial mesoderm, and it is not possible to distinguish differential expression in these tissues. The branchial arches are largely derived from neural crest, although there is also contribution from paraxial mesoderm in the core of the branchial arch in chick (37). This paraxial contribution to ventral parts of the arch is small and is confined to the core of the arch; therefore, we believe that the hybridization seen in arch 2 is largely due to expression in the neural crest. In other sections we observe expression in the superior ganglion of the IX/X ganglion complex. With the expression in the VII/VIII ganglion complex (Fig. 4A), this suggests that *Hox-2.8* is also expressed in all cranial ganglia posterior to the cutoff in the neural tube. In summary, there is no expression in the first arch at any level within the embryo, demonstrating that *Hox-2.8* expression distinguishes the second from the first arch.

Establishment of *Hox* Expression in the Surface Ectoderm

The areas of surface ectoderm lateral to the edges of the neural plate are known to produce thickenings or placodes, which generate neural derivatives (38, 39). In light of this and the recent work of Couly and Le Douarin (40) on the contributions of ectoderm lateral of the neural plate to the head, we were interested to see the extent of *Hox-2* expression in the surface ectoderm and the sequence by which it was established. Figure 3 also shows that the neural crest expressing *Hox-2.8* is located beneath surface ectoderm which does not express above background, despite the fact that extraembryonic membranes (indicated by an asterisk) express at high levels. This suggests that lack of signal in the surface ectoderm is not due to tissue thickness. At 9 days expression is starting to appear in the ectoderm that overlies the second arch crest, as shown in Fig. 4B, although it is not as yet as strong as the signal in the crest. By 9.5 days expression in the ectoderm over the second arch is at a level similar to that of the underlying crest mesenchyme. Thus expression in the surface ectoderm of the branchial arches appears later than the underlying crest in a pattern that suggests inductive interactions between the ectoderm and neural crest are required for expression in the surface ectoderm.

Summary of *Hox-2* Expression

In a similar manner we have examined the expression of other genes in the *Hox-2* complex to determine if they also had restricted patterns of expression in the branchial region. These results are summarized in Fig. 5. *Hox* genes in developing systems are thought to be one component of the process of assigning different states to otherwise equivalent groups of cells. The maintenance of a state may be manifested by the continued expression of these genes. Each branchial arch has a distinct code of *Hox-2* expression (with arch 1 not expressing any *Hox* gene), and this arch-specific *Hox-2* pattern is in the neural crest before it has reached the branchial arches. Given that *Antennapedia* class homeobox genes act as positional specifiers (10, 41–43) a specific

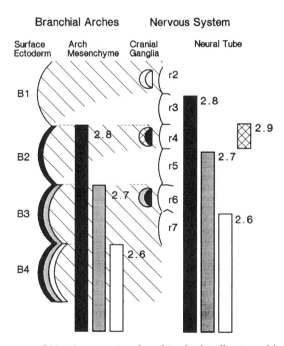

Fig. 5. Summary of *Hox-2* expression found in the hindbrain and branchial arches. The diagonal shading indicates the areas of neural plate where neural crest is produced and the branchial arch into which it migrates. The ganglion next to rhombomere 2 is the V or trigeminal, that next to rhombomere 4 is the VII/VIII or acoustic–facial complex, and those next to rhombomere 6 are the combined superior ganglia of the IX and X cranial nerves. The shading patterns shown in the cranial ganglia indicate that all the cells in a ganglion express a combination of genes and do not imply that there is spatial restriction of gene expression within a ganglion.

combination of *Hox-2* expression could provide part of the molecular mechanism for imprinting of cranial neural crest.

ROLE OF MESODERM IN NEURAL PLATE REGIONALIZATION

There is evidence to suggest that the neural induction that establishes the nervous system possesses some regional character (44, 45). Both isolated mesoderm and disaggregated mesodermal cells are able to induce neural ectoderm of a regional character in competent ectoderm. It is not clear at what resolution this induction acts and whether as discrete a set of structures as individual rhombomeres could be induced directly as a result. Recently it has been suggested that the expression of *Hox* genes seen in hindbrain is a result of a precisely spatially localized induction from the underlying mesoderm (46) which expresses a *Hox* gene in a spatially localized way in the mesoderm before expression in the ectoderm becomes apparent. However there is no evidence for the existence of a spatially localized *Hox* expression pattern in the head mesoderm underlying the hindbrain at a time before expression within the neuroepithelium is established (Fig. 2). Thus if head mesoderm is producing such a tightly localized signal, able to induce individual rhombomeres, no molecules isolated to date provide evidence for this signal.

Broad regions of the nervous system, such as midbrain versus hindbrain, or hindbrain versus spinal cord, are a result of neural induction of a spatial character. However, the refinement of this pattern may be a result of pattern-forming processes intrinsic to the neuroepithelium. Ruiz i Altalba (47) has shown that expression of *Xhox-3*, which shows regionally localized expression in parts of the hindbrain in *Xenopus* embryos, can become spatially localized in the ectoderm of total exogastrulae, in which mesoderm fails to invaginate. The precise localization of expression domains in this case cannot be due to a spatially localized signal from the mesoderm. A key event in patterning the branchial region may be the autonomous specification of regional identity in the neural plate involving the *Hox* genes, which then gives rise to the neural crest. In the head, the neural plate and its derivatives are the tissues which are regionally specified as a result of segmental processes whose final phases are intrinsic to the neuroepithelium, because they are the first tissues in the head to express *Hox-2* genes in a spatially regulated way.

TEMPORAL ONSET OF EXPRESSION

At 0 somites, when expression of *Hox-2.8*, *Hox-2.9*, and *Hox-1.6* has reached their respective anterior limits, our previous work has shown that *Hox-2.7* and *Hox-2.6* have not yet reached their anterior limits of expression (8). However, by 8.5 days post coitum (dpc) (8–12 somites) all of the genes have boundaries of expression which map to specific rhombomeres. *Hox-2.9* is unusual in that there is a dramatic change in its expression pattern between 0 and 8 somites. The early boundary which maps to r3/4 remains but posterior domains of expression decline leaving a domain restricted to r4 in the hindbrain. Also unlike other genes in the *Hox-2* complex expression recedes to background levels by 11.5 dpc. A general correlation can be drawn from our results and Fig. 6 summarizes the sequence of establishment of *Hox* gene expression in the branchial region. Successively more 5′ genes reach their anterior limit of expression at progressively later times in development. This raises the possibility that there is collinearity in the establishment or timing of anterior domains of expression with gene position in a cluster.

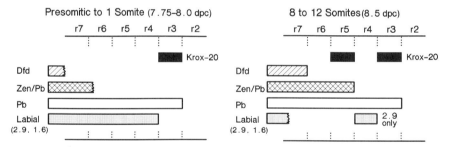

Fig. 6. Summary of the establishment of the *Hox* code in the head. The establishment of expression is shown at two stages before the morphological appearance of rhombomeres. Figure 5 shows the patterns in later stages. The presumptive rhombomere boundaries are indicated by the dotted lines. On the left-hand side of the diagram the *pb* family of genes and the *labial* family have already reached their most anterior expression boundaries. At this stage the *Dfd* and *Zen/pb* groups have not reached their anterior limits. On the right-hand side of the diagram expression of the *labial* groups begins to recede posteriorly with the exception *Hox-2.9* which is restricted to r4. Genes of all the other subfamilies have now reached their final limits, and this is the stage when neural crest migration is occurring. dpc, days post coitum

TRANSMISSION OF SPATIAL SPECIFICATION TO OTHER PARTS OF THE HEAD

Recently Couly and Le Douarin (40) investigated the fate of cells in this region of the chick body. At an early stage the surface ectoderm, prospective neural crest, and neural plate are continuous. At this time a group of marked (quail) cells were placed into the equivalent position in a chick embryo, to identify the location of their descendants and thus establish a fate map for this stage. On the basis of this it was suggested that the regions of neural tube, neural crest, and surface ectoderm that cooperate to form an arch all arise from the same axial level. Furthermore, it was suggested that all three have been initially specified as an "ectomere" on the basis of their axial position.

The expression of *Hox-2* genes in the neural tube seems to be out of phase with that of the branchial arches. The first branchial arch receives innervation from the trigeminal nerve, some of whose cell bodies are located in r3, where *Hox-2.8* is expressed (see Fig. 4A). However the rest of the first branchial arch does not express any of the *Hox-2* genes. Similarly, the second branchial arch is innervated by a nerve originating in both r4 and r5. Rhombomere 4 expresses *Hox-2.8*, while r5 expresses *Hox-2.8* and *Hox-2.7;* yet the second branchial arch does not express *Hox-2.7* in any other structures. As long as the neural tube, neural crest, and surface ectoderm each have some mechanism for specifying axial position, each component could employ a different positional signal to indicate that it is part of a particular arch. Thus the nerves and other structures of the same branchial arch need not have the same pattern of *Hox-2* expression to be able to interact with each other.

Another suggestion of the ectomere theory is that the entire ectodermal layer at this level of the body, including the presumptive epidermis, may form a genetically defined developmental unit (40) as a result of an early simultaneous specification event of neural plate, neural tube, and surface ectoderm. Yet the neural plate and neural crest express *Hox-2* genes considerably earlier than the surface ectoderm. Later, when surface ectoderm does begin to express *Hox-2* genes, it is significant that it does so after neural crest has reached the branchial arches and the pattern of expression adopted is identical to that of the crest-derived mesenchyme that underlies it.

Early specification does occur, but is confined to neural plate and presumptive neural crest. The pattern is then transferred to the branchial arches by neural crest migration, and once migration is complete, interactions occur within the branchial arch which result in the establishment of *Hox-2* expression in the surface ectoderm. The grafting experiments of Noden (27) suggest an instructive interaction between arch mesenchyme and surface ectoderm. If an arch is a genetically specified developmental unit, it is so as a result of interactions between components rather than by cooperation of units sharing the same early genetic specification.

AREAS OF REDUCED CREST EMIGRATION

Figure 3 shows that r3 does not have a population of cells lateral to the neural tube which express *Hox-2* genes. Rhombomere 5 also seems not to have hybridizing crest beside it (11), and this raises the possibility that some areas of the neural plate do not produce neural crest. Consistent with this, *Krox-20*, which is expressed in neural crest-derived boundary cap cells along the entire neuraxis, is not expressed lateral to r3 and r5 (36). SEM studies of chick and rat embryos at the time of crest emigration suggest that areas of neural tube are crest free (34, 35). A definitive proof comes from dye injections at the dorsal midline of chick neural tubes (48), which confirm that r3 and r5 do not produce any neural crest. As a result rhombomere 4 contributes neural crest to the whole of the second arch and no other. Together these data imply that the branchial arch expression that we see is out of phase with the neural tube because of the presence of crest-free rhombomeres. The lack of extensive mixing between different populations of neural crest along the rostrocaudal axis (48) would mean that during migration into an arch relative spatial positions of cells are maintained and hence pattern of gene expression. On the basis of these findings the most anterior region that produces neural crest and expresses *Hox-2*.8 would be adjacent to rhombomere 4. In a similar way the most anterior producing crest (contributing to arch 3) and expressing *Hox-2*.7 would be r6. A result of these patterns of neural crest migration would be that the neural crest in an arch expresses a *Hox-2* code related to its level of origin along the margins of the neural plate, a process summarized in Fig. 5.

MECHANISM OF HEAD SEGMENTATION

The number and size of the repeating units in neural tube and branchial arches are probably established before *Hox-2* expression reaches these regions, and the neural crest does not appear to be intrinsically segmented despite arising from a segmented structure. Experiments in amphibia involving removal of pharyngeal endoderm, which reduces the number of branchial arches, have shown that the neural crest then migrates down to fill the reduced number of arches that are available (49), suggesting that the environment is causing the neural crest to form a series of repeated structures, rather than any intrinsic property of the crest such as its pattern of gene expression. The crest-free areas described above mean that three subpopulations of neural crest with different *Hox* expression patterns are kept distinct from each other by their position of origin and subsequent migration route.

Because neural crest migrates from the neural plate, it is conceivable that by patterning the neuroepithelium, *Hox-2* genes are part of the process specifying the structures of the head and neck. However, within the branchial area it is unlikely that all spatial information, sufficient to give a detailed pattern to parts of a single arch, could be laid down within the crest before it migrates. This would require almost no cell mixing whatsoever to occur during migration, or a similarly unlikely very precise pattern of cell rearrangements. *Antennapedia* class homeobox genes are unlikely to provide information such as anteroposterior (A–P) polarity within an arch, as they are homogenously expressed there. Information in the head region for skeletal morphogenesis must also come from the crest environment. This is supported by grafts of neural plate in normal and reversed rostrocaudal orientation, in which the structures that form in the second arch are of normal rostrocaudal orientation (27).

DIFFERENCES IN EXTENT OF SPECIFICATION IN CRANIAL CREST

It is important to note that not all properties of cranial crest are consistent with regional identity being imprinted before migration. McKee and Ferguson (50) extirpated mesencephlic crest, but found no resulting facial abnormalities, as crest anterior and posterior of the lesion increased its rate of proliferation and migrated in to fill the defect. One possible interpretation is that in this experiment crest is becoming respecified, intercalating the missing positional values, although it is hard to see how the necessary communication could occur in a migratory population of cells. Alternatively this may reflect differences in properties between branchial and more anterior crest, the form of structures derived from the latter being a result of epigenetic interactions with head epithelia (32, 51). Experimental data favor the second possibility; there is little evidence to suggest differential imprinting in the crest arising from r2 and anterior, which gives rise to the upper and lower jaws and the trabeculae. When frontonasal or maxillary crest is grafted into the second arch of chick (27), it gives rise to a *mandibular* skeleton, suggesting that all anterior crest has the same positional value. The fact that no *Antennapedia* class *Hox* genes isolated to date are expressed more anterior of *Hox-2.8* in fore or midbrain suggests that other patterning systems must be operating in more anterior parts of the head. The distributions of Ankyrin and type II collagen in vertebrates, which correlate with the initial sites of endoskeletal cartilage formation, suggest a more important role for facial epithelia in determining the patterns of chondrogenesis (52). If the differences in structures formed by anterior crest are a result of interactions with the anterior cranial environment, then in the different environment of the branchial arches they form a mandible as this is some kind of "default state."

EXPERIMENTAL SUPPORT FOR THE *Hox* CODE

The correlations discussed in this paper have been generated by descriptive analysis and it is both difficult and dangerous to attempt to predict the functions of genes simply from their patterns of expression. Recently, however, direct experimental evidence has been generated to functionally link the *Hox* gene network with specification of regional identity in the vertebrate head. In two studies mutant alleles of *Hox-1* genes (1.5 and 1.6) were produced by using homologous recombination in embryonic stems cells to disrupt the endogenous genes (53, 54). Mice homozygous for these mutations were nonviable and phenotypes were primarily restricted to derivatives of the head and thorax. The neural crest was one of the major tissues affected. Several interesting observation arise from these studies. For example, different types of neural crest were altered in the mutants such that *Hox-1.6* had abnormal neurogenic crest (54) and *Hox-1.5* had defective mesenchymal crest (53). This shows that different *Hox* genes are used to differentially pattern neural crest. Another point was that not all regions which expressed the two genes were defective despite the lack of functional products in the mutants. This suggests that the genes are not required for patterning in all of their normal domains of expression and that there may be some functional overlap between *Hox* genes such that other genes compensated for the missing genes.

CONCLUSIONS

Together the molecular and embryological descriptive studies firmly suggest that one of the major embryonic contexts which utilize *Hox* genes in specifying pattern is the branchial region of the head. They suggest that the rhombomeres themselves have a role outside of the central nervous system in serving to organize the positional cues for neural crest which migrates into the branchial arches. The recent mutant experiments illustrate that these correlations are of fundamental importance during embryogenesis and will ensure that molecular pathways in head development will be an exciting area of future research.

REFERENCES

1. Graham, A., Papalopulu, N., and Krumlauf, R. (1989). *Cell* **57**, 367–378.
2. Duboule, D., and Dolle, P. (1989). *EMBO* **8**, 1497–1505.
3. Boncinelli, E., Acampora, D., Pannese, M., D'Esposito, M., Somma, R., Gaudino, G., Stornaiuolo, A., Cariero, M., Faiella, A., and Simeone, A. (1989). *Genome* **31**, 745–756.
4. Kappen, C., Schugart, K., and Ruddle, F. (1989). *Proc. Natl. Acad. Sci. U.S.A.* **86**, 5459–5463.
5. Graham, A., Papalopulu, N., Lorimer, J., McVey, J., Tuddenham, E., and Krumlauf, R. (1988). *Genes Dev.* **2**, 1424–1438.
6. Featherstone, M. S., Baron, A., Gaunt, S. J., Mattei, M. G., and Duboule, D. (1988). *Proc. Natl. Acad. Sci. U.S.A.* **85**, 4760–4764.
7. Izpisua-Belmonte, J., Falkenstein, H., Dolle, P., Renucci, A., and Duboule, D. (1991). *EMBO J.* **10**, 2279–2289.
8. Wilkinson, D., Bhatt, S., Cook, M., Boncinelli, E., and Krumlauf, R. (1989). *Nature* **341**, 405–409.
9. Lewis, E. (1978). *Nature* **276**, 565–570.
10. Akam, M. (1987). *Development* **101**, 1–22.
11. Hunt, P., Wilkinson, D., and Krumlauf, R. (1991). *Development* **112**, 43–51.
12. Hunt, P., Whiting, J., Muchamore, I., Marshall, H., and Krumlauf, R. (1991). *Development* **112** (Supplement: Molecular and Cellular Basis of Pattern Formation), 186–195.
13. Giampaolo, A., Acampora, D., Zappavigna, V., Pannese, M., D'Esposito, M., Care, A., Faiella, A., Stornaiuolo, A., Russo, G., Simeone, A., Boncinelli, E., and Peschle, C. (1989). *Differentiation* **40**, 191–197.
14. Gaunt, S. J., Sharpe, P. T., and Duboule, D. (1988). *Development* **104** (Supplement: Mechanisms of Segmentation), 169–179.

15. Dolle, P., Izpisua-Belmonte, J. C., Falkenstein, H., Renucci, A., and Duboule, D. (1989). *Nature* **342**, 767–772.
16. Izpisua-Belmonte, J.-C., Tickle, C., Dolle, P., Wolpert, L., and Duboule, D. (1991). *Nature* **350**, 585–589.
17. Nohno, T., Noji, S., Koyama, E., Ohyama, K., Myokai, F., Kuroiwa, A., Saito, T., and Tanaguchi, S. (1991). *Cell* **64**, 1197–1205.
18. Dressler, G. R., and Gruss, P. (1989). *Differentiation* **41**, 193–201.
19. Kessel, M., and Gruss, P. (1991). *Cell* **67**, 89–104.
20. Simeone, A., Acampora, D., Arcioni, L., Andrews, P. W., Boncinelli, E., and Mavilio, F. (1990). *Nature* **346**, 763–766.
21. Simeone, A., Acampora, D., Nigro, V., Faiella, A., D'Esposito, M., Stornaiuolo, A., Mavilio, F., and Boncinelli, E. (1991). *Mech. Dev.* **33**, 215–227.
22. Krumlauf, R., Papalopulu, N., Clarke, J., and Holder, N. (1991). *Sem. Dev. Biol.* **2**, 181–188.
23. Papalopulu, N., Hunt, P., Wilkinson, D., Graham, A., and Krumlauf, R. (1990). *In* "Neurology and Neurobiology," Vol. 60, "Advances in Neural Regeneration Research" (F. J. Seil, ed.), pp. 291–307. Wiley–Liss, New York.
24. Akam, M. (1989). *Cell* **57**, 347–349.
25. Balling, R., Mutter, G., Gruss, P., and Kessel, M. (1989). *Cell* **58**, 337–347.
26. Holland, P., and Hogan, B. (1988). *Genes Dev.* **2**, 773–782.
27. Noden, D. (1983). *Dev. Biol.* **96**, 144–165.
28. Verwoerd, C., and van Oostrom, C. (1979). "Advances in Anatomy, Embryology and Cell Biology," Vol. 58. "Cephalic Neural Crest and Placodes." Springer-Verlag, Berlin.
29. Nichols, D. (1981). *J. Embryol. Exp. Morphol.* **64**, 105–120.
30. Murphy, P., and Hill, R. (1991). *Development* **111**, 61–74.
31. Nichols, D. (1986). *Am. J. Anat.* **176**, 221–231.
32. Hall, B. (1987). *In* "Developmental and Evolutionary Aspects of the Neural Crest" (P. F. A. Maderson, ed.), pp. 215–259. Wiley, New York.
33. Noden, D. (1987). *In* "Developmental and Evolutionary Aspects of the Neural Crest" (P. F. A. Maderson, ed.), pp. 89–119. Wiley, New York.
34. Anderson, C., and Meier, S. (1981). *Dev. Biol.* **85**, 385–402.
35. Tan, S. S., and Morriss-Kay, G. (1985). *Cell Tissue Res.* **240**, 403–416.
36. Wilkinson, D., Bhatt, S., Chavrier, P., Bravo, R., and Charnay, P. (1989). *Nature* **337**, 461–464.
37. Noden, D. (1988). *Development* **103** (Supplement: Craniofacial Development), 121–140.
38. D'Amico-Martel, A., and Noden, D. (1983). *Am. J. Anat.* **166**, 445–468.
39. Le Douarin, N., Fontaine-Perus, J., and Couly, G. (1986). *Trends Neurosci.* **9**, 175–180.
40. Couly, G., and Le Douarin, N. (1990). *Development* **108**, 543–558.
41. Beeman, R. (1987). *Nature* **327**, 247–249.
42. Beeman, R., Stuart, J., Haas, M., and Denell, R. (1989). *Dev. Biol.* **133**, 196–209.
43. Kessel, M., Balling, R., and Gruss, P. (1990). *Cell* **61**, 301–308.
44. Saxen, L. (1989). *Int. J. Dev. Biol.* **33**, 21–48.
45. Hemmati-Brivanlou, A., Stewart, R., and Harland, R. (1990). *Science* **250**, 800–802.
46. Frohman, M., Boyle, M., and Martin, G. (1990). *Development* **110**, 589–607.
47. Ruiz i Altaba, A. (1990). *Development* **108**, 595–604.
48. Lumsden, A., and Sprawson, N., and Graham, A. (1991). *Development* **113**, 1281–1291.

49. Balinsky, B. I. (1981). "An Introduction to Embryology," 5th Ed., pp. 465–466. Saunders College, New York.
50. McKee, G., and Ferguson, M. (1984). *J. Anat.* **139**, 491–512.
51. Thorogood, P. (1988). *Development* **103**, 141–153.
52. Thorogood, P., Bee, J., and Von Der Mark, K. (1986). *Dev. Biol.* **116**, 497–509.
53. Chisaka, O., and Capecchi, M. (1991). *Nature* **350**, 472–479.
54. Lufkin, T., Dierich, A., LeMeur, M., Mark, M., and Chambon, P. (1991). *Cell* **66**, 1105–1119.

16

Murine Paired Box Containing Genes

RÜDIGER FRITSCH AND PETER GRUSS[1]
Department of Molecular Cell Biology
Max-Planck-Institute for Biophysical Chemistry
3400 Göttingen, Germany

INTRODUCTION

The key problem in developmental biology is how cells become specified and differentiate during development. It is thought that differential gene expression, controlled by transcription factors specifically expressed during development, is a key step in this process. This hypothesis was supported by the genetic and molecular analysis of *Drosophila* embryogenesis which revealed that the body plan of the fly is specified by a set of DNA-binding proteins that act as transcriptional regulators (1). Sequence comparison of *Drosophila* segmentation and homeotic genes showed that they share conserved motifs such as the homeobox (2) or the paired box (3) which turned out to encode DNA-binding domains (4, 5). Since the basic molecular mechanisms have been highly conserved during evolution, many attempts were directed toward the isolation of similar genes from other species through sequence homology to these conserved motifs. As a result, homeobox genes have been identified in a wide variety of organisms and evidence is accumulating that these genes code for developmental control factors in vertebrates as well (6–11). Similarly, paired box sequences have been identified in the genomes of different organisms such as nematode, zebrafish, frog, turtle, chick, mouse, and human (12–14). In the mouse, a family of paired box-containing (*Pax-*) genes has been isolated: another gene family with unknown function.

---

[1] Based on an evening lecture given at the P&S Biomedical Sciences Symposium, "Cell–Cell Signaling in Vertebrate Development."

Cell–Cell Signaling in
Vertebrate Development

Up to now, eight *Pax* genes, designated *Pax-1* to *Pax-8*, have been identified in the mouse genome (15). As in *Drosophila*, the mouse paired box genes sometimes contain a homeobox of the paired type. The determination of the chromosomal location of the *Pax* genes using the interspecies backcross system (16) demonstrated that they are not organized in clusters (15). Although three *Pax* genes have been located on chromosome 2, their loci are far apart. Moreover, some *Pax* genes have been mapped in the vicinity of mutants which present a phenotype coinciding with the expression pattern of the respective gene during development (see Fig. 1). Evidence will be presented demonstrating that mutations in at least two of these genes, in *Pax-1* and *Pax-6*, are responsible for the phenotypes observed.

MURINE *Pax* GENE FAMILY

Classification and Evolution

Comparison of the sequences and the genomic organization of the paired boxes led to the classification of the paired box genes into distinct classes (15). This comparison revealed that murine paired box genes have evolved through consecutive duplication events and after some divergence this resulted in the different classes which contain genes that are very similar in many respects. These paralogous genes share sequence homology inside and outside of the paired box, similar intron–exon structure, and a similar expression pattern during devel-

	chromosome	mutants	mRNA size	protein size	schematic structure PD Oct HD
Pax-1	2	*undulated* (*un*)	3.1 kb	361 aa	
Pax-2	19		4.2/4.7 kb	392/415 aa	
Pax-3	1	(*Splotch* (*Sp*)?)	3.3/3.6 kb	479 aa	
Pax-4	6		?	?	
Pax-5	4		?	?	
Pax-6	2	*Small eye* (*Sey*)	3.0 kb	422/436 aa	
Pax-7	4		4.9 kb	>300 aa	
Pax-8	2	(*Danforth's short tail* (*Sd*)?)	3.1 kb	456 aa	

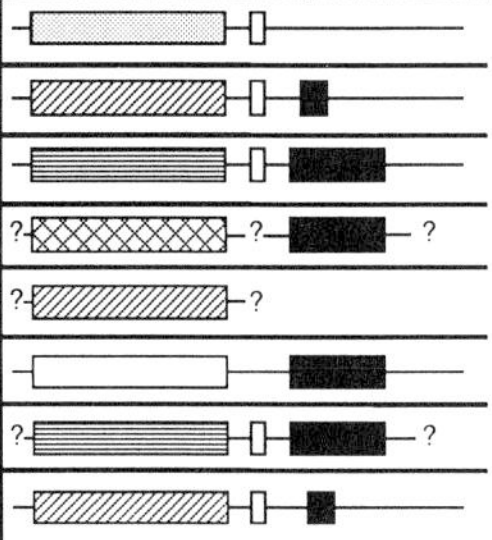

Fig. 1. Compilation of basic data for the murine *Pax* genes. aa, amino acids; kb, kilobases; HD, homeodomain; Oct, octapeptide; PD, paired domain; ?, not determined. Hatching of paired domains reflects different paired domain classes.

opment. No such paralogous genes have been found in *Drosophila.* Their occurrence in the mouse possibly reflects the higher complexity of vertebrate development. Related expression patterns observed for murine paralogous genes suggest that not only coding sequences but also at least some regulatory elements governing tissue specificity have been conserved after the duplication event. However, no *Pax* gene promoters have been analyzed so far.

Although the *Pax* genes have been cloned by multiple low-stringency hybridization screening with various probes and different cDNA and genomic libraries, it cannot be rigorously excluded that more *Pax* genes are present in the mouse genome. However, all the paired box genes that have been isolated from other vertebrate species so far have counterparts in the murine *Pax* family (14, 17–19).

Prototype Structure

Among the eight *Pax* genes, six of them, *Pax-1* to *Pax-3* and *Pax-6* to *Pax-8*, have been analyzed in detail and cDNA clones have been isolated, allowing the analysis of the proteins encoded by these genes. Parts of the putative protein sequences of *Pax-4* and *Pax-5* have been deduced from genomic sequences. Murine Pax proteins are 361 to 479 amino acids long (see Fig. 1). The prototype structure of a mouse Pax protein is as follows: the paired domain, which is 128 amino acids long, is located close to the amino terminus of the protein. It is linked by a stretch of up to 78 amino acids, which usually contains a conserved octapeptide (14), to the homeodomain of the paired type which is 61 amino acids long. The carboxy-terminal part is a sequence rich in proline, serine, and threonine, similar in amino acid composition to transcription-activating domains of transcription factors such as CTF-1 or Oct-2 (20, 21). Not all Pax proteins follow exactly this scheme; some have no (Pax-1) or only a rudimentary homeodomain (Pax-2, Pax-8), while others lack the octapeptide (Pax-4 and Pax-6). The schematic structure of the Pax proteins is shown in Fig. 1. The high content in the amino acids serine and threonine in these proteins suggests that phosphorylation is involved in the regulation of their function as demonstrated for transcription factors such as Oct-2 and Jun (21, 22).

Molecular Function

Paired domain proteins, so far analyzed, are localized in the nucleus (see *Drosophila* Pox meso and Pox neuro (23); mouse Pax-1 and Pax-2 (Fritsch and Gruss, unpublished (1991); (23a). Sequence-specific DNA binding to a fragment derived from the *Drosophila even-skipped* promoter, the e5 sequence, has been shown for the *Drosophila* paired protein (24). Binding studies revealed that the paired protein contains two independent DNA-binding domains, the homeodomain and the paired domain, that bind to different segments of the e5 sequence (5, 25). Subsequently, DNA binding to variants of the e5 sequence has been shown for all mouse Pax proteins analyzed so far (26–28). Surprisingly, despite the significant sequence divergence found within the paired domains of different classes, these domains display a similar binding specificity for at least 20 variant oligonuleotides derived from the e5 sequence. However, the significance of these *in vitro* binding experiments still remains to be demonstrated *in vivo*. It is still possible that Pax proteins of different classes bind to different target sites *in vivo*.

The paired domain contains a helix–turn–helix motif in its carboxy-terminal part. While it had been suggested that this might be a DNA-binding motif as seen for example in the homeodomain, it has been shown that the helix–turn–helix motif is not essential for DNA-binding activity (5, 27). Thus, the paired domain mediates DNA binding by a new, as yet unknown mechanism.

Binding interference analysis using different Pax proteins on various e5-derived oligonucleotides identified a core DNA sequence that was important for all Pax proteins ((27) and Chalepakis and Gruss, unpublished). *Pax-3*, which has a paired domain and a homeodomain, requires in addition an ATTA motif upstream of the core DNA sequence to stabilize the DNA-binding activity ((26) and Chalepakis and Gruss, unpublished). In contrast, Pax-1 and Pax-8, which lack a homeodomain, require additional sequences more downstream of the core DNA sequence for binding ((27) and Chalepakis and Gruss, unpublished). The DNA sequence bound by the Pax proteins is unusually large, encompassing up to 24 nucleotides in interference experiments. However, Pax proteins seem to bind as monomers, because no heterodimeric complexes could be identified in gel shift experiments where intact and truncated variants of Pax proteins were mixed ((27) and Chalepakis and Gruss, unpublished). Nevertheless, it cannot be excluded that other proteins are involved in the formation of the observed complexes.

Pax proteins are transcription-stimulating factors since they can mediate transcriptional transactivation of a minimal promoter linked to hexamers of Pax protein-binding sites ((27, 28) and Chalepakis, Fickenscher, and Gruss, unpublished). These experiments clearly demonstrate that Pax proteins can act as DNA-binding transcription factors.

Overall Expression Pattern

The murine *Pax* genes are expressed in a distinct temporal and spatial pattern during embryogenesis, starting between Day 8.0 and 9.5 postcoitum (pc). In some cases, expression remains throughout development to the adult stage. In general, genes with a homeobox appear to be expressed earlier and are restricted mainly to mitotically active cells whereas genes without homeobox seem to be involved rather in differentiating cells. Expression is generally tissue-specific. All genes are expressed in axial structures along the entire anteroposterior axis. The compilation of *Pax* gene expression domains during development is presented in Figs. 2 and 3.

Pax-1

Pax-1 was the first murine paired box gene isolated and described (12). Together with *Drosophila Pox meso* (23) and human HuP48 (14) it constitutes the class I of paired box genes. Among the mouse *Pax* genes, *Pax-1* is unique in several respects. It is the only *Pax* gene that has no counterpart in the mouse genome that is similar in sequence, structure, or expression pattern. Its basic structure seems to have evolved very early, since the *Drosophila Pox meso* gene is highly similar at least in the paired box region. HuP48 is the human *Pax-1* homologue and its protein sequence is nearly identical as far as the murine Pax-1 sequence can be compared to the HuP48 sequence.

Pax-1 is the only Pax protein completely lacking a homeodomain. In band shift and interference experiments using DNA sequences derived from the *Drosophila* e5 site, the sequence-specific DNA-binding activity of Pax-1, which is mediated by the paired domain, has been demonstrated (27). Transfection experiments with reporter constructs containing hexamers of Pax-1-binding sites showed that Pax-1 is able to transactivate an adjacent minimal promoter.

Pax-1 is also unique in that it is not expressed in the developing central nervous system. Instead, *Pax-1* is expressed in the segmented paraxial mesoderm of the developing vertebral column, the sclerotome, being later confined to the intervertebral disk anlagen and in the sternebrae of the segmented sternum (12). Expression has also been detected in unsegmented structures like the shoulder girdle, some head mesenchymal structures (Dressler, Deutsch, and Gruss, unpublished), the pharyngeal pouches I–III, and the pharyngeal pouch-derived thymus (12). A feature common to all expression domains of *Pax-1* is not apparent. However, with the exception of the thymus, all

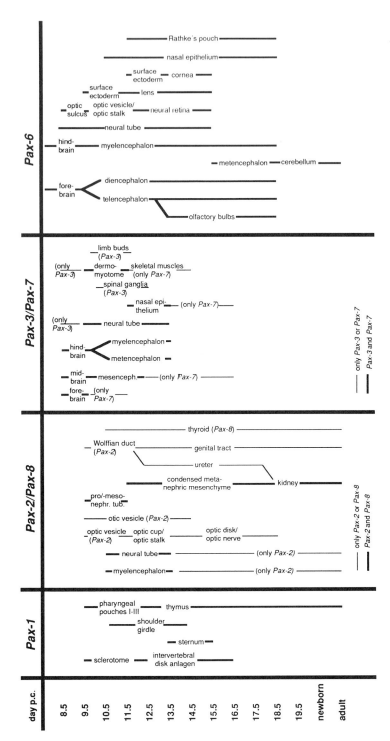

Fig. 2. Compilation of expression domains of *Pax* genes during murine development. The exact onset and offset of expression has not been determined exactly for every expression domain; this represents the expression data available thus far.

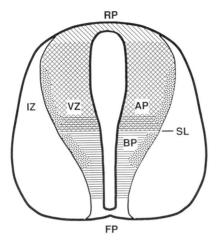

Fig. 3. Schematic representation of *Pax* gene expression in a neural tube cross section (Day 11 pc). AP, alar plate; BP, basal plate; FP, floor plate; IZ, intermediate zone; RP, roof plate; SL, sulcus limitans; VZ, ventricular zone. ◰, *Pax-3*, ◩, *Pax-7*; ▢, *Pax-2/Pax-8*; ▤, *Pax-6*..

these structures undergo chondrogenesis at some stage. Thus, *Pax-1* could be involved in some specification or early differentiation events in these tissues.

Pax-1 has been mapped to chromosome 2 with close linkage to *agouti* (13, 29, 30), locating it near the recessive mutant *undulated* (31). Homozygous *undulated* mice display malformations in the vertebral column, the sternum, and the shoulder girdle (32, 33). It has been concluded that the basic defect underlying the skeletal malformations is a reduced capacity of skeletogenic mesenchyme to condense. In particular, in the prevertebral column, this might be due to a misallocation of sclerotome cells between vertebral body and intervertebral disk anlagen (33). Sequence analysis of the *Pax-1* paired box from *undulated* mice identified a point mutation that results in a glycine-to-serine exchange within a highly conserved part of the paired domain (29). Subsequent analysis of other *undulated* alleles, named *undulated-extensive* (34) and *Undulated-short tail* (35), also revealed defects of the *Pax-1* gene (35a), thus firmly establishing the association between *Pax-1* and *undulated*. Experiments using a Pax-1 protein containing the amino acid exchange characteristic of the *undulated* mutation showed that the affinity and specificity of DNA binding as well as the capacity of transcriptional stimulation was affected, thus providing a molecular basis for the *undulated* mutation (27). The reduced DNA-binding affinity and the concomitant decreased transcriptional activation of the *undulated* Pax-1 protein are consistent with *undulated* being a recessive loss of function mutation and a hypomorph. However, a gap still remains between our understanding of the expression pattern and the molecular function on one hand and the embryological defects documented for the *undulated* mutants on the other hand.

Pax-2, Pax-5, and Pax-8

The paralogous genes *Pax-2*, *Pax-5*, and *Pax-8* constitute the class III of the paired box genes. No *Drosophila* counterpart for this class is known. A gene has been isolated from zebrafish, *pax[zf-b]*, whose gene product shows only one amino acid exchange to Pax-5 in the paired domain and displays an expression pattern that is similar to *Pax-2* (18). Thus, *pax[zf-b]* could be a Pax-5 homologue or alternatively could represent the common ancestor of *Pax-2* and *Pax-5* prior to duplication.

Since no cDNAs have been isolated for *Pax-5*, the paired box sequence has been derived from genomic clones (15). The protein sequences of Pax-2 and Pax-8 have been deduced from cDNAs (36, 37). Both proteins are similar inside and outside the paired domain (65% identity for the whole protein). Moreover, both proteins contain the conserved octapeptide. Comparison of different Pax protein sequences revealed that these proteins contain a rudimentary paired type homeodomain which encompasses 23 amino acids and includes the first α-helix (38). The functional significance of this segment is not clear. Two different types of *Pax-2* cDNAs have been isolated that represent differential splicing products. The corresponding mRNAs code for two different proteins, one harboring 23 amino acids inserted downstream of the octapeptide. This difference of 69 nucleotides, however, cannot account for the appearance of two messages of 4.2 and 4.7 kb. Thus, the molecular event underlying this difference still remains to be characterized.

Like Pax-1, Pax-2 and Pax-8 display sequence-specific DNA-binding and transcriptional activation in experiments using e5-site-derived binding sequences ((28) and Chalepakis and Gruss, unpublished). Independently, sequence-specific DNA binding has been shown for Pax-2 and Pax-8 by other groups: immunopurified Pax-2 from embryonic kidney binds e5 (23a), while Pax-8 binds to promoter sequences of the thyroid-specific genes thyroglobulin and thyroperoxidase (39a).

Pax-2 and *Pax-8* display a very similar expression pattern during embryogenesis (36, 37, 40). Only fragmentary expression data are available for *Pax-5*. *Pax-2* and *Pax-8* are expressed in the neural tube and the hindbrain, where they are restricted to a specific set of postmitotic differentiating cells on both sides of the dorsoventral midline, the sulcus limitans (see Fig. 3). While *Pax-8* is turned off at around Day 13.5 pc, *Pax-2* continues to be expressed throughout development and in adult tissue (28, 37, 40). In addition, *Pax-2* is also expressed in neural components of the developing ear and eye (40). Aside from the nervous system, both genes are also being transcribed in the developing kidney (28, 36, 37). At early stages, *Pax-2* and *Pax-8* mRNA are present in the pro- and mesonephric tubules. At later stages both genes are expressed in the condensed metanephric mesenchyme and in the resulting epithelial tubules. Additionally, *Pax-2* expression is found in the Wolffian duct and later in the ureter that buds out from the Wolffian duct. Between metanephric mesenchyme and the ureter, reciprocal induction events take place: the nephrogenic mesenchyme induces the ureter to branch whereas the branching ureter induces the mesenchyme to condense (41). Interestingly, the expression pattern of *Pax-2* and *Pax-8* coincides with these morphogenetic events. Finally, *Pax-8* is expressed in the thyroid, from the earliest stage of thyroid induction to the adult stage (37, 39a). Thus, *Pax-8* is one of the earliest markers for thyroid development.

The chromosomal localization of *Pax-2*, *Pax-5*, and *Pax-8* has been determined (15, 37). *Pax-2* is linked to several anonymous markers on chromosome 19, *Pax-5* to *Mos* and *Etl-2* (H. Neuhaus and A. Gossler, personal communication) on chromosome 4 and *Pax-8* very close to the *surfeit* locus on chromosome 2. This linkage positions *Pax-8* near the locus *Danforth's short tail* (42). This semidominant mutant is characterized by vertebral column malformations that are the result of an underlying notochord defect (43) and by a kidney defect that is intrinsic for this organ (44). Since *Pax-8* is not expressed in the notochord, one would have to assume defects (for example a deletion) affecting another, notochord-specific gene, in addition to *Pax-8*, that might be responsible for the kidney phenotype. However, no molecular data are available that support or exclude a role for *Pax-8* in this mutant.

Pax-3 and *Pax-7*

Together with the *Drosophila* genes *paired* (45), *gooseberry-distal* and *gooseberry-proximal* (46), and the human genes HuP1 and HuP2 (14), the paralogous genes *Pax-3* (26) and *Pax-7* (47) constitute class II of the paired box genes. Typically, genes of this class harbor in addition to the paired box a paired-type homeobox and, with the exception of *paired*, encode the conserved octapeptide. Remarkably, the similarities between the mouse and the *Drosophila* genes are almost exclusively restricted to the conserved boxes, excluding the possibility that they represent true homologous genes from fly and mammal. However, a gene homologous to *Pax-3*, ChPax-3, has been isolated from the chick. Both its sequence and its expression pattern are highly conserved between mouse and chick (17). Correspondingly, HuP1 and HuP2 are the human homologues of *Pax-7* and *Pax-3*, respectively.

cDNA clones have been obtained for both genes (26, 47); however, the *Pax-7* clones available do not encode the complete translated sequence. The proteins encoded by these genes are highly similar inside and outside the paired domain and paired-type homeodomain, being 85% identical as far as the sequence for the Pax-7 protein has been described. Two different *Pax-3* mRNA species have been detected by northern analysis. However, the molecular event giving rise to these different messages still has to be uncovered.

Pax-3 was the first murine Pax protein for which sequence-specific DNA binding was shown (26). Like the *Drosophila* paired protein, Pax-3 binds to variants of the *Drosophila* e5 sequence with the paired domain and the homeodomain. A Pax-7 peptide containing both the paired domain and homeodomain possesses similar DNA-binding activity (Jostes and Gruss, unpublished).

As is typical for paralogous genes, *Pax-3* and *Pax-7* exhibit a very similar expression pattern (26, 47). Both genes are expressed during early brain and neural tube development before the onset of neural differentiation. In the brain, *Pax-3* and *Pax-7* are expressed in a similar pattern in mitotically active cells in all brain vesicles. In the neural tube, both genes are restricted to the dorsal half of the ventricular zone, the alar plate, with *Pax-7* sparring the most mediodorsal part, the roof plate (see Fig. 3). *Pax-3* is expressed slightly earlier than *Pax-7*. While *Pax-3* mRNA is detectable in the neural folds at Day 8.5 pc, *Pax-7* cannot be detected before closure of the neural tube. Other expression domains that are common to both genes are the somites and subsequently the somite-derived dermomyotome. Again, *Pax-3* mRNA has been detected earlier than *Pax-7* mRNA. However, while *Pax-3* expression is turned off at around Day 11.5, *Pax-7* continues to be expressed until Day 14.5 in the myotome-derived muscles of the shoulder girdle and the trunk. In this respect, *Pax-7* could be involved in myogenic differentiation. Apart from that event, *Pax-3* and *Pax-7* are predominantly expressed in mitotically active cells. Additional expression domains of *Pax-3* have been observed in limb bud mesenchyme and neural crest derivatives such as the spinal ganglia and some cranio-facial structures.

The chromosomal location for *Pax-3* and *Pax-7* has been determined (15). *Pax-3* is closely linked to *Ctla-4*, *Vil*, and *Mylf* on chromosome 1 and Pax-7 to *Jun* and an anonymous marker on chromosome 4. Based on linkage data and expression pattern, *Pax-3* is a candidate for the semi-dominant mutation *Splotch* (48) that maps to the corresponding region. While heterozygous *Splotch* mice have defects in pigmentation, homo-zygous embryos display in addition severe defects in neural tube, brain, and spinal ganglia and die at around Day 13 pc (49). It has been concluded that the basic defects in *Splotch* resides in the dorsal neural tube and the neural crest (49). The molecular characterization of the allele *Splotch-retarded* has shown that several markers, including *Vil*, are deleted (50). Thus, it seems likely that *Pax-3* is deleted as well in this allele and contributes to the observed phenotype. However, for a clear association between *Pax-3* and *Splotch*, intragenic lesions in other alleles have to be identified. The *Splotch* mutation is discussed as a model for the human Waardenburg syndrome type I (51, 52). Thus, if defects in *Pax-3* were responsible for the *Splotch* phenotype, this human congenital disease could be analyzed experimentally in a mouse model system.

Pax-4 and *Pax-6*

Although *Pax-4* and *Pax-6* have been grouped into different classes, class IV and VI, similarities suggest that they have evolved from a common ancestor by an early duplication event and subsequent se-quence divergence (15, 38). The sequence homology between these two genes is relatively low. However, they show the same overall structure, containing both a paired box (15) and a paired type homeo-box but no octapeptide ((38) and Asano and Gruss, unpublished). Moreover, they share a similar intron–exon structure (as far as ana-lyzed) and the so-called class-specific amino acids in the conserved protein domains. Both genes are highly diverged as compared to other *Pax* genes. Genes homologous to the murine *Pax-6* gene have been isolated from zebrafish (*pax[zf-a]* (18)/*zfpax-6* (19)) and chicken (ChPax-6 (17)). Sequence and expression pattern are extremely con-served in all three species, the entire protein sequence being 97% identical between mouse and zebrafish (19).

Because no *Pax-4* cDNA clones have been obtained, the paired box and homeobox region have been isolated and sequenced from genomic clones. Although no expression data are available yet, it appears unlikely that *Pax-4* is a pseudogene, since no nonsense codon appears in the putative coding sequence and all potential splice sites are similar to known consensus sequences ((15) and Asano and Gruss, unpublished). Two types of *Pax-6* cDNAs have been isolated, representing differentially spliced messages (38). This differential splicing gives rise to a variant Pax-6 protein which harbors a 14-amino acid segment inserted into a highly conserved part of the paired domain. Thus, taking into account the previously identified elements of the paired domain important for DNA-binding activity of other Pax proteins (27), this insertion is likely to affect the DNA-binding properties of Pax-6. The DNA-binding activity of the Pax-6 protein without insertion has been demonstrated and indicated that the sequence specificity of Pax-6 differs considerably from that of other Pax protesin (Walther and Gruss, unpublished).

Pax-6 is the first *Pax* gene to be expressed during development (38). Its expression is almost exclusively restricted to the developing central nervous system. *Pax-6* mRNA has been detected in the neuroectodermal epithelia in the brain, the neural tube, the eye and the nose. No mesodermal expression has been observed for *Pax-6* and no segmented pattern typical of the other *Pax* genes. In the brain, *Pax-6* is present in all vesicles except the midbrain and in contrast to *Pax-3* and *Pax-7* is also detectable at later stages in brain development. In the neural tube, *Pax-6* expression is expressed in the ventral half of the ventricular zone, the basal plate, extending dorsal into the alar plate but sparring the most medioventral part, the floor plate (see Fig. 3). Of special interest is the expression pattern in the developing eye. *Pax-6* is expressed in the optic sulcus, in the optic vesicle and optic stalk which are derived from the optic sulcus, in the neural epithelium-like neural retina, in the surface ectoderm induced by the optic vesicle to invaginate in order to become lens, and in the surface ectoderm instructed by the lens to become cornea. Thus, *Pax-6* expression coincides with the major morphogenetic events during eye development. Remarkably, it is present in both the inducing and the responding tissues.

The chromosomal location of *Pax-4* and *Pax-6* has been determined. *Pax-4* is linked to the *Hox-1* cluster and to an anonymous marker on chromosome 6 and *Pax-6* to *agouti* and the *Hox-4* cluster on chromosome 2 (15). On chromosome 2, a semidominant mutation, *Small eye* (53, 54) is located that displays a phenotype in the eye and nose (54, 55). While heterozygous *Small eye* mice have abnormally small lens and in some cases display hydrocephaly, homozygous embryos completely fail to induce a lens, lack the nasal placodes, and die soon after birth. This phenotype coincides with the expression of *Pax-6* during development of these structures. However, no defect has been observed in the neural tube, where *Pax-6* is also expressed. Recently, analysis of different alleles of *Small eye* identified deletions and point mutations in the *Pax-6* gene (55a and Walther and Gruss, unpublished), thus providing evidence that the defects in the *Pax-6* gene are responsible for the observed phenotype. *Small eye* has been proposed as a model for human aniridia (56). Thus, in a mouse model system, this human congenital disease can now be experimentally studied.

CONCLUSIONS

In the mouse, a family of paired box-containing genes has been isolated that appears to encode important regulators of murine embryogenesis. Paired box genes have been highly conserved throughout vertebrate evolution both in sequence (>95% sequence identity between homologous proteins) and expression pattern, demonstrating that an extreme high selection pressure acts to conserve these genes and therefore suggesting an essential function. The high conservation even in zebrafish makes it likely that their molecular and embryological functions have been specified early during evolution. Since paired box sequences have been detected even in primitive invertebrates as the nematodes, one question arising is what function might have been related to the common ancestor of the paired box genes. However, the similarities between the *Drosophila* and mouse genes seem to be rather superficial and no true homologous genes between fly and vertebrates exist. Consequently, no predictions can be made concerning this point.

Although all *Pax* genes except *Pax-6* are expressed in segmented mesodermal structures like the somites or the embryonic kidney, in contrast to the paired box-containing segmentation genes in *Drosophila*, an involvement for the mouse genes in the segmentation event can be excluded since expression appears after the main segmentation processes have taken place. All *Pax* genes with the exception of *Pax-1* are expressed in the developing central nervous system. Remarkably, although segmented in some parts, as the rhombomeres of the hindbrain (57), the expression pattern of the *Pax* genes does not reflect this segmentation. In the brain, the expression domains of *Pax-3*, *Pax-6*, and *Pax-7* reside in partly overlapping, partly complementary regions as seen with many other potential regulators expressed during brain development such as Dlx (58), *En-1*, and *En-2* (59, 60) or members of the *Wnt* family (61). This pattern suggests that some *Pax* genes play a role in early regionalization of the brain. Similarily, in the neural tube, the expression domains of *Pax-3* and *Pax-7* and *Pax-6* complement each other, *Pax-3* and *Pax-7* being expressed in the dorsal half and *Pax-6* in the ventral half (see Fig. 3). This pattern points to a role in dorsoventral patterning of the neural tube of these genes. The dorsoventral polarization of the neural tube appears to be governed by the notochord which induces the formation of the floor plate that in turn itself specifies the cell types present in the ventral half of the neural tube (for review see (62)). In transplantation experiments in the chick, removal or addition of a notochord specifically altered the expression pattern of ChPax-3 and ChPax-6, suggesting that these genes at least reflect the altered polarization resulting from the manipulation (17). However, it is tempting to speculate that these genes are not only markers but are functionally involved in specifying the dorsal and ventral half by interpreting positional information directed from the notochord in an induction-like event. Expression of several other *Pax* genes coincides with morphogenetic events mediated by an inductive process, as the development of the eye (*Pax-6*) and the kidney (*Pax-2* and *Pax-8*). Remarkably, in these structures, the *Pax* genes are expressed in the inducing as well as in the responding tissues. However, to date, no functional correlation can be drawn between the expression patterns and the embryological events.

Since it has been shown that *Pax* genes encode DNA-binding transcription factors, the question arising is what do these regulators regulate. The identification of target genes will be one major step filling the gap between the expression pattern and biochemical function on one side and the embryological function on the other side. The most convincing evidence that *Pax* genes encode regulators of murine development comes from the clear association of two *Pax* genes with two classical mutations that have a phenotype resulting from a developmental defect. This clearly demonstrates that the *Pax* genes act as murine developmental control genes. The general importance of paired box genes is further highlighted by the association of at least one gene with a human congenital disease. The availability of mutants for *Pax* genes, whose phenotypes demonstrate their embryological importance, is an invaluable tool in understanding the biological processes in which *Pax* genes are involved.

ACKNOWLEDGMENTS

We are grateful to R. Balling, G. Dressler, N. Hastie, H. Neuhaus, A. Gossler, A. Püschel, S. Zannini, and R. Di Lauro for communicating results prior to publication. We thank M. Asano, G. Chalepakis, M. Goulding, B. Jostes, and especially P. Tremblay and C. Walther for helpful comments on the manuscript. This work was supported by the Max Planck Society and the Deutsche Forschungsgemeinschaft. This manuscript was prepared in October 1991.

REFERENCES

1. Ingham, P. W. (1988). *Nature* **335**, 25–34.
2. McGinnis, W., Levine, M. S., Hafen, E., Kuroiwa, A., and Gehring, W. J. (1984). *Nature* **308**, 428–433.
3. Bopp, D., Burri, M., Baumgartner, S., Frigerio, G., and Noll, M. (1986). *Cell* **47**, 1033–1040.
4. Desplan, C., Theis, J., and O'Farrell, P. H. (1985). *Nature* **318**, 630–635.
5. Treisman, J., Harris, E., and Desplan, C. (1991). *Genes Dev.* **5**, 594–604.
6. Wright, C. V. E., Cho, K. W. Y., Hardwicke, J., Collins, R. H., and De Robertis, E. M. (1989). *Cell* **59**, 81–93.
7. Kessel, M., and Gruss, P. (1990). *Science* **249**, 374–379.
8. Kessel, M., Balling, R., and Gruss, P. (1990). *Cell* **61**, 301–308.
9. Kessel, M., and Gruss, P. (1991). *Cell* **67**, 89–104.
10. Chisaka, O., and Capecchi, M. R. (1991). *Nature* **350**, 473–479.
11. Joyner, A. L., Herrup, K., Auerbach, B. A., Davis, C. A., and Rossant, J. (1991). *Science* **251**, 1239–1243.
12. Deutsch, U., Dressler, G. R., and Gruss, P. (1988). *Cell* **53**, 617–625.
13. Dressler, G. R., Deutsch, U., Balling, R., Simon, D., Guenet, J.-L., and Gruss, P. (1988). *Development* **104** (Suppl), 181–186.
14. Burri, M., Tromvoukis, Y., Bopp, D., Frigerio, G., and Noll, M. (1989). *EMBO J.* **8**, 1183–1190.
15. Walther, C., Guenet, J.-L., Simon, D., Deutsch, U., Jostes, B., Goulding, M., Plachov, D., Balling, R., and Gruss, P. (1991). *Genomics* **11**, 424–434.
16. Guenet, J.-L. (1986). *In* "Topics in Microbiology and Immunology" (M. Potter, J. Nadeau, and M. P. Cancro, eds.), Vol. 127, pp. 109–130. Springer, Berlin.
17. Goulding, M. D., Lumsden, A., and Gruss, P. (1993). *Development* **117**, 1101–1146.
18. Krauss, S., Johansen, T., Korzh, V., and Fjose, A. (1991). *Nature* **353**, 267–270.
19. Püschel, A. W., Gruss, P., and Westerfield, M. (1992). *Development* **114**, 643–651.
20. Mermod, N., O'Neill, E. A., Kelly, T. J., and Tjian, R. (1989). *Cell* **58**, 741–753.
21. Tanaka, M., and Herr, W. (1990). *Cell* **60**, 375–386.
22. Boyle, W. J., Smeal, T., Defize, L. H., Angel, P., Woodgett, J. R., Karin, M., and Hunter, T. (1991). *Cell* **64**, 573–584.
23. Bopp, D., Jamet, E., Baumgartner, S., Burri, M., and Noll, M. (1989). *EMBO J.* **8**, 3447–3457.
23a. Dressler, G. R., and Douglas, E. C. (1992). *Proc. Natl. Acad. Sci. U.S.A.* **89**, 1179–1183.
24. Hoey, T., and Levine, M. (1988). *Nature* **332**, 858–861.

25. Treisman, J., Gönczy, P., Vashishita, M., Harris, E., and Desplan, C. (1989). *Cell* **59**, 553–562.

26. Goulding, M. D., Chalepkis, G., Deutsch, U., Erselius, J. R., and Gruss, P. (1991). *EMBO J.* **10**, 1135–1147.

27. Chalepakis, G., Fritsch, R., Fickenscher, H., Deutsch, U., Goulding, M., and Gruss, P. (1991). *Cell* **66**, 873–884.

28. Fickenscher, H. R., Chalepakis, G., and Gruss, P. (1993). *DNA Cell Biol.*, in press.

29. Balling, R., Deutsch, U., and Gruss, P. (1988). *Cell* **55**, 531–535.

30. Siracusa, L. D., Silan, C. M., Justice, M. J., Mercer, J. A., Bauskin, A. R., Ben-Neriah, Y., Duboule, D., Hastie, N., Copeland, N. G., and Jenkins, N. A. (1990). *Genomics* **6**, 491–504.

31. Wright, M. E. (1947). *Heredity* **1**, 137—141.

32. Grüneberg, H. (1950). *J. Genet.* **50**, 142–173.

33. Grüneberg, H. (1954). *J. Genet.* **52**, 441–455.

34. Wallace, M. E. (1985). *J. Hered.* **76**, 271–278.

35. Blandova, Y. R., and Egorov, I. U. (1975). *Mouse News Lett.* **52**, 43.

35a. Balling, R., Lau, L. F., Dietrich, S., Wallin, J., and Gruss, P. (1992). *In* "Postimplantation Development in the Mouse," pp. 132–143. Wiley, Chichester.

36. Dressler, G. R., Deutsch, U., Chowdhury, K., Nornes, H. O., and Gruss, P. (1990). *Development* **109**, 787–795.

37. Plachov, D., Chowdhury, K., Walther, C., Simon, D., Guenet, J.-L., and Gruss, P. (1990). *Development* **110**, 643–651.

38. Walther, C., and Gruss, P. (1991). *Development* **113**, 1435–1449.

39. Gould, A. P., Brookmann, J. J., Strutt, D. I., and White, R. A. H. (1990). *Nature* **348**, 308–312.

39a. Zannini, M., Francis-Lang, H., Plachov, D., and Di Lauro, R. (1992). *Mol. Cell. Biol.* **12**, 4230–4241.

40. Nornes, H. O., Dressler, G. R., Knapik, E. W., Deutsch, U., and Gruss, P. (1990). *Development* **109**, 797–809.

41. Saxén, L. (1987). "Organogenesis of the Kidney." Cambridge Univ. Press, Cambridge.

42. Dunn, L. C., Gluecksohn-Schoenheimer, S., and Bryson, V. (1940). *J. Hered.* **31**, 343–348.

43. Grüneberg, H. (1953). *J. Genet.* **51**, 317–326.

44. Gluecksohn-Waelsch, S., and Rota, T. R. (1963). *Dev. Biol.* **7**, 432–444.

45. Frigerio, G., Burri, M., Bopp, D., Baumgartner, S., and Noll, M. (1986). *Cell* **47**, 735–746.

46. Baumgartner, S., Bopp, D., Burri, M., and Noll, M. (1987). *Genes Dev.* **1**, 1247–1267.

47. Jostes, B., Walther, C., and Gruss, P. (1991). *Mech. Dev.* **33**, 27–38.

48. Russel, W. L. (1947). *Genetics* **32**, 102.

49. Auerbach, R. (1954). *J. Exp. Zool.* **127**, 305–329.

50. Epstein, D. J., Malo, D., Vekemans, M., and Gros, P. (1991). *Genomics* **10**, 89–93.

51. Ishikiriyama, S., Tonoki, H., Shibuya, Y., Chin, C., Harado, N., Abe, K., and Niikawa, N. (1989). *Am. J. Hum. Genet.* **33**, 505–507.

52. Foy, C., Newton, V., Wellesley, D., Harris, R., and Read, A. P. (1990). *Am. J. Hum. Genet.* **46**, 1017–1023.

53. Roberts, R. C. (1967). *Genet. Res.* **9**, 121–122.

54. Hogan, B. L. M., Horsburgh, G., Cohen, J., Hetherington, C. M., Fisher, G., and Lyon, M. F. (1986). *J. Embryol. Exp. Morphol.* **97**, 95–110.

55. Hogan, B. L. M., Hirst, E. M. A., Horsburgh, G., and Hetherington, C. M. (1988). *Development* **103**(Suppl), 115–119.
55a. Hill, R. E., Favor, J., Hogan, B. L. M., Ton, C. C. T., Saunders, G. F., Hanson, I. M., Prosser, J., Jordan, T., Hastie, N. D., and van Heyningen, V. (1991). *Nature* **354**, 522–525.
56. Glaser, T., Lane, J., and Housman, D. (1990). *Science* **250**, 823–827.
57. Lumsden, A. (1990). *Sem. Dev. Biol.* **1**, 117–125.
58. Price, M., Lemaistre, M., Pischetola, M., Di Lauro, R., and Duboule, D. (1991). *Nature* **351**, 748–751.
59. Davis, C. A., and Joyner, A. L. (1988). *Genes Dev.* **2**, 1736–1744.
60. Davis, C. A., Noble-Topham, S. E., Rossant, J., and Joyner, A. L. (1988). *Genes Dev.* **2**, 1736–1744.
61. Roelink, H., and Nusse, R. (1991). *Genes Dev.* **5**, 381–388.
62. Lumsden, A. (1991). *Cell* **64**, 471–473.

17

Gene Regulation during Nerve Growth Factor-Induced Differentiation

EDWARD B. ZIFF[1]
Howard Hughes Medical Institute
Department of Biochemistry
Mental Health Clinical Research Center
Center for Neural Science
New York University Medical Center
New York, New York 10016

We have studied the program of neuronal differentiation induced in PC12 cells by nerve growth factor (NGF) (Metz *et al.*, 1988). PC12 is a cell line derived from a pheochromocytoma of the rat which expresses receptors for NGF and epidermal growth factor (EGF) (Greene and Tischler, 1976, 1982). Treatment with NGF induces a phenotypic change from that of a chromaffin cell to that resembling a sympathetic neuron (Unsicker *et al.*, 1978; Aloe and Levi-Montalcini, 1979; Anderson and Axel, 1986). Treatment with EGF induces cell viability but not the neuronal phenotypic changes (Huff and Guroff, 1979; Boonstra *et al.*, 1983).

Here we review studies by our laboratory of gene regulation induced by NGF during PC12 cell differentiation. We have previously identified three kinetically distinct classes of genes induced by NGF. These include (1) the immediate early genes, induced within 10 to 15 min of NGF or EGF treatment (Greenberg *et al.*, 1985), (2) the delayed early genes, induced 1 to 2 hr after NGF stimulation, and (3) the late genes which are induced approximately 18 to 24 hr following NGF stimulation (Leonard *et al.*, 1987). Among the immediate early

---

[1] Based on the closing address given at the P&S Biomedical Sciences Symposium, "Cell–Cell Signaling in Vertebrate Development."

Cell–Cell Signaling in
Vertebrate Development

genes are the c-*fos* gene, which encodes c-Fos, a basic-region leucine zipper protein which forms a complex with c-Jun that binds to specific sequences in DNA, so-called "AP-1 sites" (reviewed by Angel and Karin, 1991). The gene encoding tyrosine hydroxylase (TH, tyrosine monooxygenase), the enzyme which carries out the rate-limiting step in catecholamine biosynthesis, falls in the delayed early class (Gizang-Ginsberg and Ziff, 1990). The gene encoding peripherin (Thompson and Ziff, 1989), which encodes a peripheral neuron-specific intermediate filament protein (Leonard *et al.*, 1988), falls in the late class (Leonard *et al.*, 1987; Thompson *et al.*, 1992).

EXPRESSION OF IMMEDIATE EARLY GENES

Immediate early genes including c-*fos* are induced by protein synthesis-independent mechanisms (Greenberg *et al.*, 1986b). This implies that all components of the signal transduction system necessary for immediate early gene activation preexist in the cell prior to induction. The c-*fos* promoter is inducible by a wide range of agents which transmit signals across the plasma membrane, including many growth factors, phorbol esters, agents which elevate the level of cyclic AMP, and certain neurotransmitters (reviewed by Angel and Karin, 1991; Sheng and Greenberg, 1990). cAMP-dependent activation takes place through several elements. One of these is a binding site for the CREB protein located at residue -60 (Sassone-Corsi *et al.*, 1988). The major target for activation by growth factors is the serum regulatory element (SRE) (Treisman, 1985), a region which binds a multitude of transcription factors and which is centered at approximately residue -300. A schematic diagram of protein factors binding to the SRE is given in Fig. 1. Within the SRE is located a palindromic sequence called the dyad symmetry element (DSE). Within the DSE lies the motif $CC(A/T)_6GG$, called the CArG box, which is the binding site for serum regulatory factor (SRF) (Treisman, 1986; Gilman *et al.*, 1986; Norman *et al.*, 1988).

The importance of SRF is emphasized by the fact that mutations which prevent its binding also greatly decrease the responsiveness of the c-*fos* promoter to growth factor-induced signals (Greenberg *et al.*, 1987). Also placement of a DSE sequence upstream from a surrogate gene, such as the β-globin gene, confers growth factor responsiveness (Siegfried and Ziff, 1989). It is not yet established how SRF responds to growth factor-induced signals in activating transcription from the c-*fos* promoter.

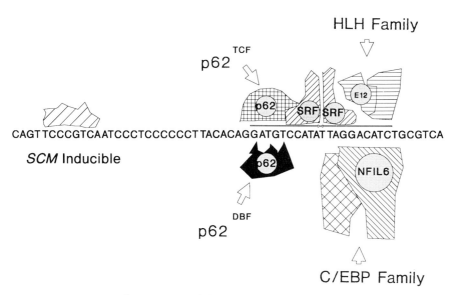

Fig. 1. Serum regulatory region of the c-*fos* gene. The DNA-binding sites for transcription factors which interact with the SRE are indicated and described in the text.

A possible clue to the mechanism of growth factor activation comes from studies of yeast mating-type-specific gene regulation. Genes whose expression is mating-type-specific contain within their regulatory regions binding sites for the factor MCM1, a factor also known as PRTF or GRM (for a review see Andrews and Herskowitz, 1990). The MCM1 DNA-binding site sequence is homologous to the CArG box (Hayes *et al.*, 1988). Cloning and sequencing of the c-*fos* regulatory factor SRF revealed that the protein is substantially homologous to MCM1 within its DNA-binding domain (Norman *et al.*, 1988). The protein kinase Raf has been implicated in the mechanism which conveys the signal from the cytoplasm to the c-*fos* promoter (Jamal and Ziff, 1990; Siegfried and Ziff, 1990). MCM1 controls gene activity by cooperation with positive- and negative-acting factors that are mating-type-specific in their expression. Yeast cells with mating-type α express the coactivator α1 which binds adjacent to MCM1 and induces the expression of genes which are specifically active in α cells. The repressive factor α2 binds both immediately upstream and downstream from MCM1 in the promoters of genes that are specifically repressed in α cells. The similarities between SRF and MCM1 and their binding sites (Hayes *et al.*, 1988) suggest that SRF operates through interaction with positive- and negative-acting factors that bind to adjacent sites.

While SRF binds to the CArG box within the DSE, other factors which may cooperate with SRF bind at closely neighboring positions. These include p62^{TCF}, a protein which forms a ternary complex through contacts with both SRF and DNA (Shaw *et al.*, 1989). p62^{TCF} has been implicated in the response of the c-*fos* gene to phorbol (esters) (Graham and Gilman, 1991) and has recently been shown to be a member of the ETS family of transcription factors (Hipskind *et al.*, 1991). Our laboratory has identified two additional transcription factors not previously recognized to interact with the SRE. One of these is the rat homolog of the factor NFIL-6 (Metz and Ziff, 1991a,b), also known by several other names (Akira *et al.*, 1990; Chang *et al.*, 1990; Descombes *et al.*, 1990) including c-EBP-β (Cao *et al.*, 1991). rNFIL-6 is a basic-region leucine zipper transcription factor distantly related to c-Fos which belongs to a family containing additional members c-/EBP-α and c/EBP-γ (Cao *et al.*, 1991). The different family members are highly homologous within their DNA-binding domains but differ elsewhere in their structures. Expression of the C/EBP family is regulated in a characteristic manner during cell differentiation, for example, in adipocytes (Cao *et al.*, 1991). The DNA-binding site for rNFIL-6 in the SRE is a consensus C/EBP-binding sequence which overlaps the CArG box and binding site for SRF (Metz and Ziff, 1991a). We have also shown that the helix–loop–helix protein E12 binds immediately adjacent the rNFIL-6 site to an "E-box" sequence, whose canonical structure is CANNTG (Metz and Ziff, 1991b). E12 is a ubiquitously expressed protein which forms heterodimers with other helix–loop–helix proteins. These binding partners, in contrast to E12, are tissue-specific in their expression. Such partners include MyoD, a muscle cell-specific factor (Lassar *et al.*, 1989), and MASH1 and MASH2, which are expressed in PC12 cells (Johnson *et al.*, 1990). The heterodimeric complexes of helix–loop–helix proteins have E-box DNA-binding activity.

The fact that at least two factors binding at positions adjacent to SRF, rNFIL-6 and E12, belong to transcription factor families which are tissue-specific in their expression raises the possibility that the c-*fos* gene is tissue-specific in its responses.

Our studies of rNFIL-6 indicate that this factor is under control by cAMP (Metz and Ziff, 1991a). In unstimulated PC12 cells, rNFIL-6 is distributed between the cytoplasm and the nucleus. However when these cells are treated with forskolin, which elevates cyclic AMP levels, the rNFIL-6 protein concentrates in the nucleus. It also undergoes phosphorylation and can be detected in complexes with 60-kDa and 42-kDa proteins, the latter of which is also phosphorylated. The movement of this factor to the nucleus and its phosphorylation are both independent of new protein synthesis. rNFIL-6 has been shown to be capable of transactivating transcription through the DSE; however, the relationship of its action to SRF binding is not yet known. The binding sites for SRF and rNFIL-6 overlap extensively and it is not yet established if both factors can bind simultaneously to the same DNA molecule. Preliminary studies indicate that immunologically related protein can be detected in neurons in rat spinal ganglia and in purkinje cells of the adult rat (J. Gorham, R. Metz, H. Baker, and E. Ziff, unpublished).

EXPRESSION OF DELAYED EARLY GENES

The tyrosine monooxygenase gene (TH gene) falls within the delayed early class (Gizang-Ginsberg and Ziff, 1990). Its expression is induced by NGF and EGF at the transcriptional level, with peak activity approximately 1 hr following NGF stimulation as shown in Fig. 2 for NGF. After that time, TH transcription rapidly declines. The promoter region of the TH gene contains a sequence closely related to the binding site for the Fos–Jun complex, an AP-1 site. We have shown that the proteins in nuclear extracts of PC12 cells that associate with the AP-1 site of the TH promoter change following NGF treatment (Gizang-Ginsberg and Ziff, 1990). In untreated cells, very little protein associates with this element. However following NGF stimulation, protein factors binding to the site increase dramatically in parallel with the activation of TH transcription. Antibody competition studies show that c-Fos is a component of this complex. Its presence in extracts from NGF-stimulated PC12 cells has been confirmed by Western blot analysis. If the Ap-1 site of the TH promoter is mutated so that the Fos–Jun complexes can no longer bind, the promoter loses its responsiveness to NGF. This demonstrates that interaction of the Fos–Jun complex with the TH promoter is required for gene induction. Surprisingly, at the time when TH transcription declines, PC12 cell nuclear factors which can associate with the AP-1 site of the TH promoter remain abundant. However, Western blotting revealed that in fact there is a change in these protein components which occurs as transcription is repressed. The c-Fos protein, which accumulates at early times when the gene is active, soon decays and is replaced by antigenically related members

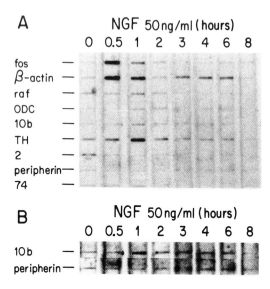

Fig. 2. Induction of transcription by NGF in PC12 cells. Analysis of gene transcription by the nuclear runoff assay is shown. Expression of the tyrosine monooxygenase (TH) and the 10b gene follows expression of c-*fos* and actin genes (Gizang-Ginsberg and Ziff, 1990). (A, B) Two different exposures of the same assay (see text).

of the c-Fos family including FosB (Gizang-Ginsberg and Ziff, unpublished observations). Fos-related proteins are also capable of interacting with the Jun protein (or other members of the Jun protein family) to form heterodimers which can bind to AP-1 sites. This suggests a model given in Fig. 3 in which complexes containing the c-Fos protein are activators of transcription while Fos-related proteins that predominate at times the promoter is shut off contribute to a complex in which TH gene activity is repressed.

Our results establish that TH transcription can be indirectly regulated by NGF through the agency of the immediate early c-Fos protein. Because the c-*fos* gene may be induced by neurotransmitters (reviewed by Sheng and Greenberg, 1990), transcription of the TH gene is potentially under control of transsynaptic stimuli. Because the TH enzyme catalyzes the rate-limiting step in the production of the catecholamine family of neurotransmitters, this control mechanism potentially allows incoming transsynaptic signals to regulate the production of neurotransmitters which in turn stimulate other neurons.

Such a mechanism could allow a neuron to express TH enzyme in proportion to the level of its excitation. Baker and co-workers (Stone *et*

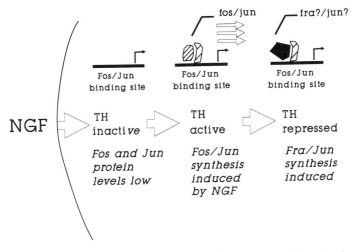

Fig. 3. Model for the control of TH expression by members of the c-Fos family. The TH promoter contains a DNA element, the TH-FSE (Gizang-Ginsberg and Ziff, 1990) which binds the c-Fos–Jun complex and is required for activation of TH transcription. Stimulation of c-Fos expression induces transcription through this element. As time progresses, levels of c-Fos decline and Fos-related proteins accumulate. The latter are proposed to be repressors of TH transcription.

al., 1990) have studied the expression of the TH gene in interneurons in the olfactory bulb. They have shown that expression of the TH enzyme is dependent on odorant-induced synaptic input from olfactory neurons. The precise mechanism of this control is not established but it potentially relies on activation of the c-*fos* gene, as demonstrated here in PC12 cells.

EXPRESSION OF LATE GENES

The peripherin gene encodes a neuron-specific intermediate filament protein structurally related to desmin and vimentin (Leonard *et al.*, 1988). The peripherin gene is induced by NGF in PC12 cells at approximately the time that NGF-dependent morphological changes become evident (Leonard *et al.*, 1987). These changes include extension of neurites. Interestingly, peripherin is localized in neurite extensions of PC12 as well as in cell bodies (Gorham *et al.*, 1991). It is found in axons of mature neurons of the peripheral nervous system (Gorham *et al.*, 1991; Escurat *et al.*, 1990). Although the function of peripherin and other intermediate filaments is not yet established, it is tempting to speculate that peripherin contributes to either neurite extension or neurite function. These roles would be in agreement with the coincidence of its expression with the onset of the NGF-induced PC12 cell morphological changes.

Restriction of expression of peripherin to mature, postmitotic peripheral neurons (Gorham et al., 1991) indicates that in vivo, induction of transcription of the peripherin gene is somehow coupled to cell-cycle exit. A functional analysis of the peripherin promoter has revealed positive and negative regulatory elements (Thompson et al., 1992). However, there is no evident AP-1-binding site for the Fos–Jun family of transcription factors. This is in agreement with the fact that c-Fos protein has decayed from the cell by the time expression of peripherin commences and the fact that peripherin expression appears to be coupled to cell-cycle exit. It is also noteworthy that while EGF and NGF both induce immediate early c-fos gene expression (Greenberg et al., 1985) and delayed early TH gene expression (Gizang-Ginsberg and Ziff, 1990, and unpublished), only NGF induces morphological differentiation and expression of the peripherin gene. This suggests that during the early phase of the response to NGF and EGF, second messenger signals generated by NGF and EGF which regulate gene activity are similar in their effects. However, they diverge afterward such that EGF induces a proliferative response while NGF induces a differentiation response characterized by cell-cycle exit. The mechanisms which enable NGF to halt PC12 cell proliferation and induce PC12 cell differentiation are not known but are central to its function in nervous system development.

PROTO-ONCOGENES AND NEURAL DEVELOPMENT AND FUNCTION

Tumors arise when genes which encode cell-cycle regulatory functions mutate and are no longer capable of governing cell proliferation and cell-cycle exit. A large component of cell cycle control is provided by extracellular signals conveyed to cells by receptors for growth factors which span the plasma membrane and are activated by ligand binding. With the EGF receptor and the trk receptor for NGF, the cytoplasmic domains are protein tyrosine kinases. Activation of protein kinase function leads to a cascade of second messenger production and the communication of signals to the nucleus which activate early response gene function (as described above). Mutations which deregulate this signaling pathway may alter growth factor production, receptor function, cytoplasmic signaling, and nuclear proto-oncoprotein production. Such mutations most frequently act in a dominant fashion in their deregulation signal transduction. One step in cell transformation is thought to occur when this pathway is constitutively active. The family of tumor suppressor proteins, encoded by recessive proto-oncogenes, limits the action of the dominant pathway. Among the candidates for such regulators is the recessive proto-oncogene product, the retinoblastoma protein (Rb protein) and a related 107-kDa protein (reviewed by Buchkovich *et al.*, 1990).

One may assess the effects of dominant mutations of signal transducing proteins in cultured cells by introducing a plasmid which expresses the dominant-acting mutant oncogenic protein. However, the realities of genetic analysis of cultured animal cells make a similar approach to investigation of recessive oncoprotein function difficult. Fortunately, the DNA tumor viruses encode transforming proteins, including the human adenovirus E1a protein, which bind the Rb and 107-kDa proteins (for a review see Buchkovich *et al.*, 1990). Binding is thought to block the capacities of the cellular tumor suppressor proteins to limit cell proliferation. Therefore, introduction into cells of genes encoding DNA tumor virus-transforming protein such as E1a provides a means for disrupting recessive oncoprotein function. We have introduced the adenovirus E1a gene into PC12 cells by selection for co-transformation to neomycin resistance. Cotransformant PC12 cells which express E1a exhibit a dramatically altered phenotype (Boulukos and Ziff, 1993). This includes a change from round, weakly adherent and slow-growing cell type to a highly adherent, flat, rapidly growing cell. The cells are also unresponsive to NGF and EGF, most likely as a consequence of failure to express receptors for these factors. Although the c-*fos* gene remains inducible, the TH and peripherin genes are not expressed. These findings suggest a dependence of the neuronal phenotype on Rb or other tumor suppressor protein whose function is disrupted by E1a. It is intriguing to speculate that one target of an NGF-induced second messenger is a member of the family of tumor suppressor proteins. Disruption of the function of such a protein would disable NGF-dependent cell-cycle exit and terminal differentiation. This possibility is supported by the normal physiological role of tumor suppressor proteins in limiting cell-cycle progression.

Control of other agents, such as the Myc protein, is also likely to be important in the final steps of neuronal differentiation. Myc is a nuclear phosphoprotein which is expressed at high levels in proliferating cells and a number of tumor types and has many properties of a transcription factor (reviewed by Prendergast and Ziff, 1992). It is induced in PC12 cells by NGF and EGF (Greenberg *et al.*, 1985). Down-regulation of Myc expression is a feature of terminal differentiation of many cell types. Control of *myc* gene expression or function most likely is required during the terminal stages of neuronal differentiation as well. If indeed Myc is a transcription factor, it may be necessary to shut off the expression of genes under Myc control in order to achieve orderly cell-cycle exit and terminal neuronal differentiation.

SUMMARY

These studies reveal an elaborate program of gene regulation induced in PC12 cells by growth factors which is summarized in part in Fig. 4. Although the mitogenic factor EGF and the differentiation factor NGF appear to be similar in their actions during the first several hours of stimulation, NGF evokes a specialized program leading ultimately to neuronal differentiation which may rely on tumor suppressor "antioncogenic" factors to evoke cell-cycle exit. Although functional neurons are postmitotic, the early response genes may still be induced in these cells by transmembrane signals. Particularly noteworthy is the

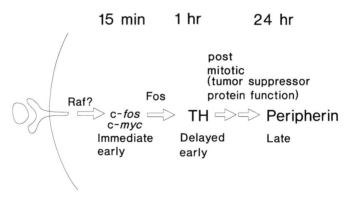

Fig. 4. Model for NGF-induced differentiation of PC12 cells. NGF induces transcription of immediate early genes inducing c-*fos* and c-*myc*. The c-Fos protein is a regulator of the delayed early tyrosine hydroxylase gene. An unknown mechanism induced by NGF perhaps employing tumor suppressor proteins drives the cells from active cycling enabling the expression of late genes including that for the intermediate filament protein peripherin.

increasing evidence that neurotransmitters can induce early response gene expression (Greenberg *et al.*, 1986a) in mature, postmitotic neurons. This suggests that growth simulatory and differentiation-inducing growth factors may employ the same signaling transduction pathways as neurotransmitters to control gene activity. If indeed this is the case, studies of the mechanisms which allow growth regulatory pathways to convey environmental signals may also be relevant to understanding how neurons respond to their environment. Such responses may provide plasticity or control homeostatic processes such as synthesis of neurotransmitters or their receptors.

ACKNOWLEDGMENTS

I thank T. Serra for encouragement and aid in preparation of the manuscript. Edward B. Ziff is an Investigator of the Howard Hughes Medical Institute.

REFERENCES

Akira, S. H., Isshiki, T., Sugita, O., Yanabe, S., Konoshita, Y., Nishio, T., Nakajima, T., Hirano, and Kishimoto, T. (1990). A nuclear factor for IL-6 expression (NF-IL6) is a member of a C/EBP family. *EMBO J.* **9**, 1897–1906.

Aloe, L., and Levi-Montalcini, R. (1979). Nerve growth factor-induced chromaffin cells *in vivo* into sympathetic neurons: Effect of anti-serum to nerve growth factor. *Proc. Natl. Acad. Sci. U.S.A.* **76**, 1246–1250.

Anderson, D. J., and Axel, R. A. (1986). A bipotential neuroendocrine precursor whose choice of cell rate is determined by NGF and glucocorticoids. *Cell* **47**, 1079–1090.

Andrews, B. J., and Herskowitz, I. (1990). Regulation of cell cycle-dependent gene expression in yeast. *J. Biol. Chem.* **265**, 14057–14060.

Angel, P., and Karin, M. (1991). The role of Jun, Fos and the AP-1 complex in cell-proliferation and transformation. *Biochim. Biophys. Acta* **1072**, 129–157.

Boonstra, J., Moolenaar, W. H., Harrison, P. H., Moed, P., van der Saag, P. T., and de Laat, S. W. (1983). Ionic responses and growth stimulation induced by nerve growth factor and epidermal growth factor in rat pheochromocytoma (PC12) cells. *J. Cell. Biol.* **97**, 92–98.

Boulukos, K. E., and Ziff, E. B. (1993). Adenovirus 5 E1a proteins disrupt the neuronal phenotype and growth factor responsiveness of PC12 cells by a conserved region 1-dependent mechanism. *Oncogene* **8**, 237–248.

Buchkovich, K., Dyson, N., Whyte, P., and Harlow, E. (1990). Cellular proteins that are targets for transformation by DNA tumor viruses. *Ciba Found. Symp.* **150**, 262–271.

Cao, Z., Umek, R. M., and McKnight, S. L. (1991). Regulated expression of three C/EBP isoforms during adipose conversion of 3T3-L1 cells. *Genes Dev.* **5**, 1538–1552.

Chang, C. J., Chen, T. T., Lei, H. Y., Chen, D. S., and Lee, S. C. (1990). Molecular cloning of a transcription factor, AGP/EBP that belongs to members of the C/EBP family. *Mol. Cell. Biol.* **10**, 6642–6653.

Descombes, P., Chojkier, M., Lichtsteiner, E., Falvey, E., and Schibler, U. (1990). LAP, a novel member of the C/EBP gene family, encodes a liver-enriched transcriptional activator protein. *Genes Dev.* **3**, 1541–1551.

Escurat, M., Djabali, K., Gumpel, M., Gros, F., and Portier, M. M. (1990). Differential expression of two neuronal intermediate-filament proteins, peripherin and the low-molecular-mass neurofilament protein (NF-L), during the development of the rat. *J. Neurosci.* **10**, 764–784.

Gilman, M. Z., Wilson, R. N., and Weinberg, R. A. (1986). Multiple protein-binding sites in the 5' flanking region regulate c-fos expression. *Mol. Cell. Biol.* **6**, 4305–4316.

Gizang-Ginsberg, E., and Ziff, E. B. (1990). Nerve growth factor regulates tyrosine hydroxylase gene transcription through a nucleoprotein complex that contains c-Fos. *Genes Dev.* **4**, 477–491.

Gorham, J. D., Baker, H., and Ziff, E. B. (1991). Differential spatial and temporal expression of two type III intermediate filament proteins in olfactory receptor neurons. *Neuron* **7**, 485–497.

Graham, R., and Gilman, M. (1991). Distinct protein targets for signals acting at the c-fos serum response element. *Science* **251**, 189–192.

Greenberg, M. E., Greene, L. A., and Ziff, E. B. (1985). Nerve growth factor and epidermal growth factor induce rapid transient changes proto-oncogene transcription in PC12 cells. *J. Biol. Chem.* **260**, 14101–14110.

Greenberg, M. E., Greene, L. A., and Ziff, E. B. (1986a). Stimulation of neuronal acetylcholine receptors induces rapid gene transcription. *Science* **234**, 80–83.

Greenberg, M. E., Hermanowski, A. L., and Ziff, E. B. (1986b). Effect of protein synthesis inhibitors on growth factor activation of c-fos, c-myc and actin gene transcription. *Mol. Cell. Biol.* **6**, 1050–1057.

Greenberg, M. E., Siegfried, Z., and Ziff, E. B. (1987). Mutation of c-fos gene dyad symmetry element inhibits serum inducibility of transcription *in vivo* and nuclear regulatory factor binding *in vitro*. *Mol. Cell. Biol.* **7**, 1217–1225.

Greene, L. A., and Tischler, A. S. (1976). Establishment of a nonadrenergic clonal line of rat adrenal pheochromocytoma cells which respond to nerve growth factor. *Proc. Natl. Acad. Sci. U.S.A.* **23**, 2424–2428.

Greene, L. A., and A. S. Tischler (1982). PC12 pheochromocytoma cultures in neurobiological research. *Adv. Cell. Neurobiol.* **3**, 2424–2428.

Hayes, T. E., Sengupta, P., and Cochran, B. H. (1988). The human c-fos serum response factor and the yeast factors GRM/PRTF have related DNA-binding specificities. *Genes Dev.* **2**, 1713–1722.

Hipskind, R. A., Rao, V. N., Mueller, C. G., Reddy, E. S., and Nordheim, A. (1991). Ets-related protein Elk-1 is homologous to the c-fos regulatory factor p62TCF. *Nature* **354**, 531–534.

Huff, K. R., and Guroff, G. (1979). Nerve growth factor induced reduction in epidermal growth factor responsiveness and epidermal growth factor receptor in PC12 cells: An aspect of cell differentiation. *Biochem. Biophys. Res. Commun.* **89**, 175–180.

Jamal, S., and Ziff, E. B. (1990). Transactivation of the c-fos and β actin genes by v-raf suggests a mechanism for early response gene control by transmembrane signals. *Nature* **344**, 463–466.

Johnson, J. E., Birren, S. J., and Anderson, D. J. (1990). Two rat homologues of Drosophila achaete-scute specifically expressed in neuronal precursors. *Nature* **346**, 858–861.

Lassar, A. B., Buskin, J. N., Lockshon, D., Davis, R. L., Apone, S., Hauschka, S. D., and Weintraub, H. (1989). MyoD is a sequence-specific DNA binding protein requiring a region of myc homology to bind to the muscle creatine kinase enhancer. *Cell* **58**, 823–831.

Leonard, D. G. B., Ziff, E. B., and Greene, L. A. (1987). The detection and characterization of messenger RNAs regulated by nerve growth factor in PC12 cells. *Mol. Cell. Biol.* **7**, 3156–3167.

Leonard, D. G., Gorham, J. D., Cole, P., Greene, L. A., and Ziff, E. B. (1988). A nerve growth factor-regulated messenger RNA encodes a new intermediate filament protein. *J. Cell Biol.* **106**, 181–193.

Metz, R., Gorham, J., Siegfried, Z., Leonard, D., Gizang-Ginsberg, E., Thompson, M., Lawe, D., Kouzarides, T., Vosatka, R., MacGregor, D., Jamal, S., Greenberg, M. E., and Ziff, E. B. (1988). Gene regulation by growth factors. *Cold Spring Harbor Symp. Quant. Biol.* **53**, 727–737.

Metz, R., and Ziff, E. B. (1991a). Cyclic AMP stimulates the C/EBP-related transcription factor rNFIL-6 to translocate to the nucleus and induce c-fos transcription. *Genes Dev.* **5**, 1754–1766.

Metz, R., and Ziff, E. B. (1991b). The Helix–Loop–Helix protein E12 and the C/EBP related factor NFIL-6 bind to neighboring sites within the c-fos serum response element. *Oncogene* **6**, 2165–2178.

Norman, C., Runswick, M., Pollock, R., and Treisman, R. (1988). Isolation and properties of cDNA clones encoding SRF, a transcription factor that binds to the c-fos serum response element. *Cell* **55**, 989–1003.

Prendergast, G., and Ziff, E. B. (1992). A new bind for myc. *Trends Genet.* **8**, 91–96.

Sassone-Corsi, P., Visvader, J., Ferland, L., Mellon, P. L., and Verma, I. M. (1988). Induction of proto-oncogene fos transcription through the adenylate cyclase pathway: Characterization of a cAMP-responsive element. *Genes Dev.* **2**, 1529–1538.

Shaw, P. E., Schroter, H., and Nordheim, A. (1989). The ability of a ternary complex to form over the serum response element correlates with serum inducibility of the human c-fos promoter. *Cell* **56**, 563–572.

Sheng, M., and Greenberg, M. E. (1990). The regulation and function of c-fos and other immediate early genes in the nervous system. *Neuron* **4**, 477–485.

Siegfried, Z., and Ziff, E. B. (1989). Transcriptional activation by serum, PDGF, and TPA through the c-*fos* DSE: Cell type specific requirements for induction. *Oncogene* **4**, 3–11.

Siegfried, Z., and Ziff, E. B. (1990). Altered transcriptional activity of c-*fos* promoter plasmids in v-*raf* transformed NIH 3T3 cells. *Mol. Cell. Biol.* **10**, 6073–6078.

Stone, D. M., Wessel, T., Joh, T. H., and Baker, H. (1990). Decrease in tyrosine hydroxylase, but not aromatic L-amino acid decarboxylase, messenger RNA in rat olfactory bulb following neonatal, unilateral odor deprivation. *Brain Res. Mol. Brain Res.* **8**, 291–300.

Thompson, M. A., and Ziff, E. B. (1989). Structure of the gene encoding peripherin, an NGF-regulated neuronal specific type III intermediate filament protein. *Neuron* **2**, 1043–1053.

Thompson, M. A., Lee, E., Lawe, D., Gizuna-Ginsberg, E., and Ziff, E. B. (1992). NGF-induced derepression of peripherin gene expression is associated with alterations in proteins binding to a negative regulatory element. *Mol. Cell. Biol.* **12**, 2501–2513.

Treisman, R. (1985). Transient accumulation of c-*fos* RNA following, serum stimulation requires a conserved 5' element and c-fos 3' sequences. *Cell* **42**, 889–902.

Treisman, R. (1986). Identification of a protein-binding site that mediates transcriptional response of the c-*fos* gene to serum factors. *Cell* **46**, 567–574.

Unsicker, K., Kirsch, B., Otter, U., and Thoenen, H. (1978). Nerve growth factor-induced fiber outgrowth from isolated rat adrenal chromaffin cells: impairment by glucocorticoids. *Proc. Natl. Acad. Sci. U.S.A.* **75**, 3498–3502.

Index

261

D

E